LET THE GOOD TIMES ROLL

To Dad
April 26 2017
Happy Birthday
Burt

OTHER BOOKS BY ROBERT FRITCHEY

WETLAND RIDERS

MISSING REDFISH

LET THE GOOD TIMES ROLL

Louisiana Cashes in Its Chips with the 1995 Net Ban

ROBERT FRITCHEY

NEW
MOON

GOLDEN MEADOW, LOUISIANA

Copyright © 2017 by Robert Fritchey

All rights reserved

Published by New Moon Press

www.newmoonpress.com

Edited by Nancy Schoeffler
Cover Design by Kathleen Joffrion
Line drawings by Lee Roy Tooke
All other images by author unless noted

ISBN 978-0-9963882-9-0
Library of Congress Catalog Card Number: 2016913843

Printed in the United States of America

For information regarding special discounts
for bulk purchases, please contact:
www.newmoonpress.com

For everyone who did what they could

PREFACE

The Gulf of Mexico is a mediterranean sea, landlocked but for the Yucatan Channel and Florida Straits. Warm Caribbean currents flood into this basin through the Channel and the overflow jets out the Straits and up the Atlantic Coast as the Gulf Stream.

The Gulf is also fed by rivers that enter from Mexico and the five American states on its northern shoreline. Rich in nutrients, their fresh inflow lingers in indentations along the coast, and barrier islands further impede mixing with the salty sea. Such protected estuaries, with their softened blend of fresh and salt water, serve as nurseries for an abundance of seafood species.

Oysters, clams, crustaceans such as shrimp and crabs, and a myriad of finfish thrive there.

Native Americans naturally harvested this bounty, and coastal tribes had little need to venture inland. Later, the Spanish and then the French colonized parts of the coast, giving the Gulf a Mediterranean flavor that was soon diluted by a steady influx of English and Scotch-Irish from the American states.

While the earliest settlers harvested seafood for their own subsistence, by the late 19th century the American Gulf Coast from Key West to Brownsville was dotted with coastal communities whose economies depended almost exclusively on renewable marine resources. Such centers of commerce attracted successive waves of workers from around the world including the Canadian Maritimes,

LET THE GOOD TIMES ROLL

China, the Philippines, the Canary Islands, Italy, Greece, Scandinavia, Croatia, Vietnam, Cambodia and Mexico.

Many of these immigrants came from fishing communities, and brought harvesting and processing techniques that advanced the development of the Gulf's fishing economy.

In the earliest days, all of the fishing occurred near shore. By the 1880s, a fleet of Pensacola-based schooners was sailing to the Campeche Bank, off the Yucatan, to handline red snapper. From such beginnings, the offshore industry evolved into that of today, with far-ranging diesel-powered vessels that bring in tuna, swordfish, grouper and other deepwater fish, and steel-hulled shrimp trawlers massive enough to be called "slabs."

Through it all, the smaller-scaled artisanal fleet continued to develop in the shallows along the Gulf's edge. From nets of cotton or linen, and boats driven by wind or muscle, the inshoremen progressed to speedy inboard- or outboard-powered skiffs, and webbing of modern synthetics. Likewise, the buyers and sellers of fish continually probed new markets for the fishermen's catch, and in so doing nourished seafood consumers locally, nationally and internationally while steadily bolstering this sustainable industry's contribution to local, state and national economies.

As the 20th century drew to a close, a convergence of forces virtually eliminated this industry.

Wild seafood species multiply free of charge, without planting, fertilizing, irrigating, weeding or spraying. The annual crops need only to be harvested yet the waters don't willingly yield their bounty and the arduous work leaves little in reserve for the type of tending that the fish do need.

PREFACE

To sustain their abundance, coastal fish require clean, free-flowing waters and lush natural habitats. These simple needs are threatened by a plague of societal impacts. And as society grows, so do the threats.

Unable to lobby or vote, the fish are best represented by those who depend most directly upon them. It's an easy fit for commercial fishermen because, as unabashed natural predators, they require the same pure waters and natural landscapes as their prey.

Yet, even while the fishing industry steadily grew in value, it was outstripped in economic and political influence by newer coastal enterprises: From the western Gulf, spreading east, there was the oil business, and from the eastern Gulf, spreading west, tourism.

Like the fishermen, the oilmen labored to produce a unique product, export it from the coast, and bring in money. But without the limits inherent to sustainable endeavors, the mining of fossil fuels mushroomed to such a gargantuan scale that it miniaturized commercial fishing's political and economic importance, while the natural incompatibility of oil and water weighted fish and fishermen with additional stresses from pollution and habitat destruction.

For tourism boosters, the Gulf Coast proved an easy sell though west of Florida sandy playgrounds were limited. The Gulf's primary selling points were its temperate winter climate and the same abundant resources that built the commercial fishery.

Recreational angling began to take off during the plush Gilded Age of the late 19th and early 20th centuries, as did the production of fishing gear, books and magazines related to that outdoor activity. Tracking the growth in the nation's population and economy, the sport—and the industry supporting it—suffered downturns during World War I and the Great Depression, grew steadily during the 1930s and began to boom amid the post-World War II prosperity, as evidenced by the explosion in fishing tackle sales, from $35 million in 1939 to $130 million in 1947.

From the 1960s through the 1990s, as the nation's 78 million Baby Boomers came of age, the number of sport fishermen shot through the roof.

Commercial fishermen may not have welcomed the newcomers with open arms, but they did not begrudge them their fair share of the resource. The naturally competitive sportsmen covetously eyed the seafood harvesters' catch and needed little prodding by recreational media and other industry members to lobby for increasing shares of this publicly owned resource; as they did, the sportsmen consistently portrayed the traditional food providers as pillagers and themselves as "conservationists," an effective yet cynical strategy considering that, contrary to public perception, the recreationists' cumulative harvest had in many cases grown to far exceed that of the commercial harvesters.

Nonetheless, the sportsmen's chorus was mysteriously amplified in the early 1990s when virtually every environmentally themed nonprofit in the country joined in an unprecedented environmental education campaign which left the general public convinced that every fish everywhere was endangered by commercial fishing.

With that advantage, and in an orchestrated effort, sport-fishing interests around the country mobilized to ban the use of the most ancient and essential of the coastal fishermen's harvesting tools—the net.

Away from the Gulf of Mexico, commercial fishermen and their allies rebuffed the sportsmen's efforts in every state but one. The distant fights—in Alaska, Oregon, Washington, Pennsylvania, New

PREFACE

Jersey and North Carolina—all revolved around some coveted trophy fish.

On the fertile Gulf it was the wetlands-loving redfish that stirred the passions of coastal anglers and led to the seminal 1980s fish fights in Texas, where sportsmen rallied behind elite leaders to monopolize the species as a "gamefish," then went on to banish nets from that oil state's waters. Encouraged by those victories, and in collaboration with the same commanders, like-minded anglers across the Gulf later waged their own campaigns, which culminated in the mid-1990s net-ban battles in Florida, Alabama, Mississippi and Louisiana.

In their campaigns, recreational advocates made outlandish claims which I recognized because I'd supported myself for a time by fishing with nets. So, with a passable ability to craft sentences, I took on the task of writing a book that would help to set the record straight. In time, it became apparent that the completion of such a comprehensive state-by-state effort would be too far into the future. So I broke it down into a collection of separate books, the "Gulf Wars" series.

"Missing Redfish," a history of the red drum's management on the Gulf, was the first work in that series; "Let the Good Times Roll" is the second.

Because net fishing is a folk occupation, handed down over the generations from fisherman to apprentice, banning the use of nets is a one-way train that virtually eliminates not only the ability to produce seafood but a traditional and sustainable way of life. These wars on the Gulf, therefore, marked the end of an age and the beginning of another in nearly every state.

Some of the changes associated with this transition can be readily quantified, such as the plummeting production of seafood. Losses to

LET THE GOOD TIMES ROLL

our coastal heritage are more difficult to measure, so each of the historical narratives is supplemented with a few profiles of people who fish.

As for the long-term effects on the actual health of the Gulf Coast's environment and fishery resources, only time will tell. One certainty is that those resources will be worth far less to society; another is that their actual status will be more difficult to assess minus the ongoing interaction of professional harvesters.

Even though I'd fished for years, commercially and recreationally, I learned much in writing this series—not just about the Gulf of Mexico's ecology and fisheries, but how mass movements are organized, and how man relates to himself, his fellows, and to nature. It's often not a pretty picture.

"Sustainable use" is simply the practice of continuing to do what we've always done, as active participants in the natural world. According to its proponents, if you don't use living renewable resources you lose them, as their value to society diminishes. As we occupy the driver's seat of our own evolution that's worth considering. Otherwise, our ready tendency to believe in the paucity of those resources may prove self-fulfilling.

Louisiana, which in the 1990s offered a cornucopia of seafood, could in time prove a case in point.

—Robert Fritchey
Golden Meadow, Louisiana

CONTENTS

• *Introduction* • 1

• *Part One* •
LET THE GOOD TIMES ROLL

CHAPTER 1
SEAFOOD FACTORY 29

CHAPTER 2
RALLYING THE TROOPS 43

CHAPTER 3
REDFISH: BONE OF CONTENTION 53

CHAPTER 4
CONVINCING THE PEOPLE 65

CHAPTER 5
THE HIGH ROAD 103

CHAPTER 6
IN THE LEGISLATURE 115

CONTENTS, CONTINUED

CHAPTER 7
NOT OVER YET *167*

CHAPTER 8
PARTING SNAPSHOTS: POUNDS AND DOLLARS *189*

• *Part Two* •
THREE LOUISIANA FISHERFOLK

CLIFF GLOCKNER, LACOMBE *203*

KERRY LEBAUVE, COCODRIE *227*

ROBERT FRITCHEY, GOLDEN MEADOW & LEEVILLE *263*

CONTENTS, CONTINUED

• *Appendix* • 301

TOOLS OF THE TRADE 303
A Glossary of Some Fishing Nets

TABLE 1 308
Louisiana Commercial and Recreational
Spotted Seatrout Landings, in Pounds, 1990-2014

TABLE 2 309
Commercial Landings of Gulf of Mexico Red Drum,
in Pounds, 1887-2014

TABLE 3 313
Recreational Landings of Gulf of Mexico Red Drum,
in Pounds, 1981-2013

TABLE 4 315
Recreational Landings of Gulf of Mexico Red Drum,
in Numbers of Fish, 1979-2013

TABLE 5 317
Number of Recreational Fishing Trips in Gulf of Mexico,
For All Species, 1981-2013

BIBLIOGRAPHY 319

Market bound. Louisiana net fishermen returning from a successful trip for pompano around the Chandeleur Islands. *(Brian Gauvin photograph)*

INTRODUCTION

Most of Louisiana's shoreline on the Gulf of Mexico is bordered by a half-sunk prairie that provides food and shelter for marine life and helps generate some of the nation's greatest fisheries.

These renewable fishery resources are said to be held in "public trust," which means in essence that they're owned by everyone. And when everyone shares something valuable, there's always somebody who thinks they're entitled to more than anyone else.

So it was with Louisiana's coastal finfish: "Let the Good Times Roll" is the story of how the fish were won, or lost, depending on your perspective.

Commercial fishermen are well acquainted with the threats that a runaway recreational fishery poses to their way of life. For other readers, the next few pages are intended to provide enough background to get up to speed on this issue, beginning with a few basic terms:

"Commercial fishing" is so named because the fish that commercial fishermen catch are sold and thus enter commerce. This enables commercial fishermen to earn their livelihoods, and consumers to purchase their catch in markets and restaurants.

LET THE GOOD TIMES ROLL

"Recreational fishing" is so named because recreational fishermen fish for sport, not profit. Sport fishermen are often called "anglers," a reference to the curve or angle within their hooks. Recreational fishermen, who typically earn their livelihoods in some way unrelated to fishing, may keep some fish for their personal use but are usually not permitted to sell their catch.

Commercial and recreational fishermen both support sizeable industries, one geared toward the distribution and consumption of fish, as food, and the other based mostly on the consumption of goods and services used to catch fish for fun.

Commercial fishermen run their family businesses as efficiently as possible, and try to keep their expenses to a minimum. The goods that they do consume include the mostly no-frills industrial hardware in their boats, engines, nets and other equipment, and the fuel and galley supplies they and their deckhands use up on their fishing trips.

Commercial fishermen sell their catch to dockside buyers, whose expenses include labor, packaging, refrigeration and ice-making equipment, extreme utility bills, and trucks that move the fishermen's harvest up the marketing chain to urban distributors who endure similar costs as they process and distribute fresh fish to markets and restaurants.

Recreational fishermen don't need to turn a profit on their fishing, and spend their discretionary income on a variety of sport-fishing related goods including tackle, boats, engines, vehicles; on their trips to the coast their expenses include fuel, bait, food and drink, and lodging. The sportsmen outnumber commercial fishermen thousands to one, and their cumulative expenditures are of a scale to warrant the attention of corporate manufacturers and retailers, and boosters of coastal tourism.

INTRODUCTION

A third "user group" is a hybrid of the other two: fishing guides, or charter-boat captains, earn their livelihoods by being paid to take recreational anglers fishing. Like the private sport fishermen who fish on their own, those in the for-hire sector are generally prohibited from selling their catch for food.

Fishery biologists include the government and university scientists who estimate fish populations and the appropriate surplus that can be harvested while leaving enough in the water to sustain the species.

Fishery management is the science, art, and the political process that regulates the people who fish, to ensure that they catch neither too many nor too few. It's not easy, on a good day, with the fish being out of sight and their populations swinging wildly in response to the ongoing changes within their natural environment. Competition between the different sectors makes it even more challenging.

"Allocation" is the subset of fishery management that determines who gets what.

Fishery resources may be allocated by a variety of means, yet common sense suggests that they be distributed in a manner that's equitable enough to ensure the continued participation of the different user groups. But when reason takes a back seat to emotion, anything can happen.

TEXAS: REDFISH GAMEFISH, THEN NETS

The Gulf Coast Conservation Association spearheaded the mid-1990s campaign to ban commercial netting in Louisiana. Headquartered in Houston, the self-proclaimed "conservation" group was first organized in the late 1970s by a group of Texas sport fishermen to "protect" the state's spotted seatrout and redfish from commercial fishermen. (Commonly known on the Gulf as the "redfish," that species is more properly referred to by scientists as the "red drum.")

These two fish were the backbone of both the sport and commercial fisheries in the nearshore waters of Texas, as well as every other Gulf Coast state including Louisiana.

Commercial landings of redfish and trout had been recorded annually since the late 1880s. Reliable landings data on the sport fishery didn't become available until nearly a century later, when the federal government commenced its annual marine recreational fishing survey in the early 1980s. Even so, it was common knowledge among biologists across the Gulf that the total recreational take of both species far surpassed that of the commercial harvesters by the late 1970s if not well before.

Until then, regulation of the redfish and trout fisheries had been virtually non-existent. And with comparatively few participants, using relatively inefficient equipment in highly productive waters, nor was it needed.

But when it became apparent—at least in the Gulf's more populated "growth" states of Florida and Texas—that too many redfish were being removed from the system, the Lone Star State was the first to address the problem.

The Red Drum Conservation Act of 1977 marked Texas' first foray into management of this popular species. A compromise between the sport and commercial fishing industries—which were then still more or less evenly matched in political clout—the act instituted a first-ever annual quota on the commercial harvest and first-ever daily limits on the number of redfish that sportsmen could retain.

The law included several additional constraints on the harvest of redfish, mostly by commercial fishermen. Had their effects on the

INTRODUCTION

fish's population been monitored and tweaked over time, both industries might have been sustained indefinitely. Leaders of the Gulf Coast Conservation Association, however, had no interest in sharing. They wanted exclusive access to the redfish as well as the trout, and made no bones about it.

GCCA was still a fledgling in 1977, not yet ready for the sort of campaign that dominated a legislative session. Over the next few years, while the harvesting restrictions eroded the finances of the commercial industry, the sport-fishing group matured into a lobbying powerhouse with a well-funded political action committee, full-time executive director, staff, and some 7,000 dues-paying members as well as 30,000 affiliate members in other sportsmen's groups. By 1981, GCCA was ready to go for gamefish.

"Gamefish" is an official designation that prohibits the sale of a fish species, thereby rendering it off limits to the commercial sector.

To help convince the Texas Legislature to declare both redfish and trout gamefish, GCCA's leadership reached out to the state's resource management agency for assistance. But when a legislator in GCCA's camp directed fishery biologists within the Texas Parks & Wildlife Department to draw up the enabling legislation, they refused. There was no scientific basis to justify re-allocating the species exclusively to the recreational sector, they said.

Science was the rudder that steered fishery management, and the job of fishery biologists was to provide the truest data they could on fish populations. The surest way for a professional biologist's data—and reputation—to be called into question was to take sides in an allocation dispute. Biologists of integrity therefore strictly adhered to the flat mantra, "A dead fish is a dead fish," because, whether it died in the meshes of a gill net or on a billionaire's gold-plated hook, the loss to the population was identical.

Nonetheless, Texas Parks and Wildlife soon began to advance a perspective of the fishery that was based less on scientific truth and more on the belief of GCCA's upper echelon that commercial fishing was a scourge, and their own activities blameless. The agency's turnaround followed GCCA's takeover of the governor-appointed, six-member Texas Parks & Wildlife Commission, which directed policy within the state agency.

After the promotion of an avowed gamefish proponent to head the department's fishery division, Texas Parks and Wildlife abandoned any pretense of neutrality on the issue and joined GCCA as a partner in its campaign to allocate the state's redfish and trout to the anglers.

The daily catches of commercial net fishermen were typically quite a bit larger than those of individual sport fishermen. Not only could those large hauls cause resentment among anglers—particularly after a less-than-satisfying day on the water—they fueled the perception that the overall landings by seafood producers far outweighed those of recreational fishermen. In reality, with these contested species, at least, the relatively small catches by relatively numerous anglers added up to a pile of fish—at least three-fourths of the trout and redfish taken each year.

That inconvenient truth hindered GCCA's efforts to convince fair-minded legislators to take the remaining allocation from commercial fishermen and consumers. To compensate for the lack of a sound biological justification for GCCA's gamefish proposal, Parks & Wildlife, with the sport-fishing group, presented the politicians with a compelling economic argument.

INTRODUCTION

While most, if not all, publicly employed natural resource economists maintained that a fishery's contribution to the economy was maximized by sharing it among the various user groups, Texas Parks and Wildlife officials testified that the fish should be allocated exclusively to the recreational fishery because the anglers spent more money in pursuit of their hobby than commercial fishermen earned by producing food. GCCA buttressed that economic argument with generous campaign contributions to the state's politicians, and in 1981 the Texas Legislature declared both redfish and spotted seatrout gamefish.

The two species were popular in Texas restaurants and had traditionally been the inshore commercial fishermen's bread and butter. But fishermen learned that they could stay in business by netting a variety of other species, many of which were of little interest to recreational fishermen. Then, in 1988, the Texas Parks & Wildlife Commission, which had never had a representative from the commercial fishing industry, simply banned the use of nets by proclamation.

Commercial fishermen fell back on trotlines—strings with hooks attached—which limited their catches to even fewer species, mostly black drum and sheepshead.

In no other Gulf Coast state would a net ban be so easily accomplished as it was in Texas, where the government had granted the authority to make that decision to a single panel of sportsmen.

As the Gulf Coast Conservation Association exported its gamefish/net ban agenda across the Gulf, it would find some friends among the enforcers of fishery laws—whose jobs naturally became easier with less gear in the water—but it would never again enjoy the backing of an entire agency that, like a ventriloquist's dummy, spouted information in support of the group's initiatives.

Texas Parks & Wildlife itself continued to assist GCCA in its expansion. Besides making its personnel available to testify in other states, the department continued to disseminate information that reinforced the group's campaign rhetoric, which generally equated commercial fishing with depletion, even criminality, and touted gamefish and net bans as the first steps back to the good old days of plenty.

In a June 5, 2006, press release, marking the death of a commissioner who'd led the effort to pass the 1981 gamefish bill, TP&WD claimed that the law "took redfish and seatrout from *commercially overfished* species on the brink of collapse to the premier recreational catch on the Texas Coast." (Author's italics.)

As recently as 2011, in its *Texas Parks and Wildlife Magazine*, the agency claimed that the redfish population in the 1970s had been decimated by "illegal gill netting. Monofilament gill nets in the hands of a few commercial fishing outlaws were devastatingly effective" and could "wipe out an entire school of reds in a single night. The scene played itself over and over again, taking a collectively greater toll on the species, until recreational fishermen brought the issue to a head...."

Passage of the gamefish bill in 1981 "paved the path for a dramatic recovery," said Parks & Wildlife, which barely acknowledged that "progressively tightened recreational fishing regulations" also played a part in that "dramatic recovery."

INTRODUCTION

LOUISIANA: REDFISH GAMEFISH

By the 1990s, redfish were undergoing similarly dramatic recoveries in all the other Gulf Coast states, after a new fishery revealed that the management of their own sport and commercial fisheries needed to be tightened as well.

Red drum had a uniquely resilient life cycle: They spent their first few years in the states' shallow coastal waters, then became sexually mature and migrated offshore where they lived for decades. Each year they returned to the beaches to spawn.

Unlike the smaller inshore juveniles, which had traditionally been harvested for the market, the much larger and older spawners—called "bull reds"—had little commercial value. That is, until the mid-1980s when New Orleans Chef Paul Prudhomme's blackened redfish recipe made the big fish more palatable.

Found in great schools offshore, the bulls were targeted with purse seines, massive nets that were worked from much larger vessels than the pickup-sized boats traditionally used by the inshore netters. The new fishery drew a lot of attention and was soon interrupted; because it occurred mostly in offshore waters—under federal rather than state jurisdiction—the federal government coordinated the research to determine sustainable harvest rates.

With the overview provided by that research, scientists were surprised to learn that the number of juvenile fish escaping each state's inshore waters, to join the brood stock, had plummeted back in the 1970s. Consequently, the offshore population was in decline. To rebuild it, federal managers in the late 1980s pressured the states to restrict their sport and commercial fisheries to allow a minimum targeted percentage of each annual crop of young fish to survive long enough to escape offshore.

As Louisiana's lawmakers instituted the cutbacks, GCCA—already the experienced veteran of the Texas gamefish war as well as a couple other scaled-down but just as successful gamefish campaigns in Alabama and South Carolina—convinced them to declare the redfish a gamefish "temporarily."

The red drum had been the wintertime staple for the Bayou State's inshore fishermen: Not only were other seafood species like shrimp and crabs scarce at that time of year but the low winter tides and cold water temperatures forced the reds from the shallows into deeper water where they could more easily be netted. As in Texas, their loss crippled the commercial industry while the recreational lobby only grew stronger.

WHO'S IN CHARGE?

Louisiana was the third state, after Texas and Alabama (1982) to start its own GCCA chapter. Not long after it was organized, in 1984, its executive director described the Texas-based group as a pyramid organization: "Everybody takes their orders from one man," he said.

Walter Fondren, of Houston, was a founder and served as CCA's chairman until his death in early 2010. (As GCCA extended its reach to the Atlantic and Pacific coasts, it was renamed the Coastal Conservation Association (CCA) in 1997; on the Gulf Coast both names—and their acronyms—remained in use.) Fondren was often described as an Exxon heir because his grandfather was one of the founders of Humble Oil, which merged with Standard Oil of New Jersey in 1972 to form that corporation. A burly man who at one time considered going into professional football, he was active in fishery management in Texas. And as a recreational fishing representative on

INTRODUCTION

the Gulf of Mexico Fishery Management Council, from 1982-1992, he influenced federal fishery policy and networked with like-minded anglers across the Gulf who wished to replicate his group's gamefish success in their own states.

Over the years, Fondren and the other anglers on GCCA's board of directors allowed its salaried staff of mostly public relations and communications professionals to assume a larger role in running the group.

As they worked full-time to advance GCCA's objectives, staff members learned to augment the not inconsiderable resources of its angling membership with the support of industrial interests that also stood to benefit from a diminished commercial fishery.

With the growing assistance of outdoor-oriented media, sportfishing guides, retailers of sporting goods, corporate manufacturers and distributors of tackle, boats, motors and trailers, as well as their national trade entities such as the American Sportfishing Association, the "conservation" group CCA became indistinguishable from an advocacy group for the recreational industry. And there wasn't anyone in the recreational fishing industry who didn't want to attract as many people as possible to the coast.

The problem was that a deeply embedded fishing economy and culture already existed there, and they were both sustained by the use of nets.

NETS

Nets enabled commercial fishermen to produce enough fish to earn a living and support the other businesses that depended on their harvest. But the same concentrated production that nets made possible also aroused primal feelings in sport fishermen, who frequently voiced their disdain for the equipment with the vacuous pronouncement, "A net is a net is a net."

In fact, there were many types of nets, which commercial fishermen deployed in many different ways. (In the Appendix, some nets that were used in coastal Louisiana are illustrated in "Tools of the Trade.") In political or educational campaigns, however, a simple message was best. So "gill nets" became the enemy.

More efficient than other nets for the capture of most fish, gill nets were indeed the most widely used. But from the perspective of the savvy anti-net warriors, who relied on manipulative words and images to overcome the lack of any scientific justification for their initiatives, the worst—or best—thing about a gill net was its name.

With its hard "g" and semblance to "kill," there was nothing cuddly about the word "gill."

As for the other types of nets in use at the time, the "trammel" in "trammel net" was an interesting word that sounded more agreeable than "gill," while the "seine" in "haul seine" was not only sonorous to the ear but suggestive of the river that meandered through romantic Paris.

So trammel nets and haul seines were rarely if ever mentioned in the sportsmen's ban-the-net campaigns. Yet when gill nets were outlawed, these other nets—which had special applications that allowed fishermen to harvest certain species that couldn't be taken with gill nets—went right along with them.

Because they restrained fish as they tried to swim through their uniformly sized meshes, gill nets were highly discriminating as to the size of fish they captured. That selectivity enabled managers to govern the size and age of fish caught by commercial fishermen, simply by

INTRODUCTION

regulating the mesh size of their nets.

In addition to mesh size, nets were also regulated by length and material composition, and restricted further as to when and where they could be set.

Though usually regarded as commercial gear, gill nets were used

Fishery biologists routinely use gill nets in their work. For example, in the Chesapeake Bay's watershed, the Atlantic sturgeon was believed to have been on the verge of extinction until recently when researchers—sampling with large-meshed anchored gill nets—discovered that the species was far more abundant than previously thought. In above photo, Dr. Matt Balazic (right) of Virginia Commonwealth University's Rice Center checks a sturgeon caught in a gill net in the tidal James River in September, 2013. He is being assisted by Chuck Frederickson, formerly the Lower James Riverkeeper. After tagging with acoustic tags, researchers released this fish and the others they netted, unharmed, then tracked their movements to learn more about the location and timing of their spawning.
(Leslie Middleton photo courtesy of Chesapeake Bay Journal.)

by some recreational fishermen as well. State agency biologists also routinely deployed gill nets—and trammel nets and seines—to sample and monitor fish populations.

The term "gill net" referred to the manner in which the webbing itself restrained fish. The construction of the entire nets, which included their corklines and leadlines, were tailored to the target species and conditions in which they were to be used. "Stab nets," for instance, were gill nets weighted with enough lead to overcome the buoyancy of the corkline, which allowed them to sink to the bottom of deeper waters. There were "flag nets," which had no leadline at all. And window nets, tied-down nets—the list went on. Though gill nets could be constructed in any number of ways, they were deployed in just two, either actively or passively.

A runaround gill net, or "strike net," as it was called on the Gulf Coast, was used actively: When a fisherman made a strike, he ran the net out around a school of fish, then quickly hauled it back onboard and moved on.

Passively, gill nets could be run out and drifted—with the fisherman's boat tied to one end—or they could be anchored in a fixed location, like a spider's web, as "set nets."

Historically, Louisiana's commercial fishermen had actively fished with trammel nets and seines. After gill nets were introduced to the state, in the 1960s and 1970s, most finfishermen adopted them as their gear of choice.

Initially, gill nets were worked actively as well; because runaround gill nets required both specialized equipment and experience to handle, the fishery was limited to professional fishermen. In large bodies of water, such as the inland Lake Pontchartrain, professional fishermen also anchored out nets and, mostly, tended them regularly.

INTRODUCTION

In Louisiana's "open access" fishery, however, anyone who forked out about $100 in license fees could legally net and sell wild fish; many folks did and though considered to be commercial they weren't necessarily as responsible or committed as full-time fishermen.

Short set nets proliferated in the same shallow and accessible waters that were traversed by sport fishermen, who resented having to navigate around the bobbing corklines. That the gear was often left untended further rankled anglers. And though their biological impact was more perceptual than actual, set nets developed into a major source of contention as the number of sport fishermen grew.

For the most part, anglers in Louisiana were completely oblivious to the use of actively fished strike nets, which fishermen carried in their boats until they were briefly deployed. With little actual knowledge of the fishery, the state's sport fishermen considered the conspicuous "set nets" to be synonymous with all "gill nets," which the Texas-based GCCA promised to eradicate.

VERBAL AMMUNITION

"With public sentiment, nothing can fail; without it, nothing can succeed," said Abraham Lincoln.

To popularize their cause and isolate the commercial fishery from potential supporters, GCCA's operatives repeated the same negative catchwords that had proven effective in previous campaigns. Gill nets thus became "walls of death" whose "greedy" operators "might as well be using dynamite" because their gear was "indiscriminate," which practically spelled out "criminal."

In addition to those old standards, the sportsmen's arsenal of verbal ammunition was bolstered in the early 1990s when a handful of charitable foundations began to indoctrinate the public about its fisheries.

LET THE GOOD TIMES ROLL

Federal fishery management was conducted by regional panels such as the Gulf of Mexico Fishery Management Council which was comprised of scientists and knowledgeable sport and commercial fishermen, as specified in the Magnuson Act, the 1976 law that standardized how the nation's fisheries were to be managed in federal waters. (Federal waters extended 200 miles beyond state waters, which in Louisiana reached 3 miles off the coast.)

The Magnuson Act was periodically reauthorized, which allowed it to be amended by Congress. In anticipation of the act's scheduled reauthorization, in the early to mid-1990s, the Pew Charitable Trusts and a few other moneyed foundations assembled the Marine Fish Conservation Network, a coalition of mostly environmental and recreational fishing organizations, to tighten various provisions in the law.

To win public support for its proposed reforms, the Network permeated the media with tailored messages that had been developed with the assistance of public relations professionals, and further refined by testing with polls and focus groups.

To disenfranchise commercial fishermen on the federal management panels, they were characterized as "foxes in the hen house." Fishing became no longer fishing but an "extractive" activity, differentiating it from "non-consumptive" uses such as "ecotourism." "Industrial" fishing conjured an image of out-of-scale, overly mechanized and technological harvesting, which struck a negative chord with the public.

In an elementary public relations ploy, environmental campaigners linked "commercial" with "extinction," until fish populations that actually numbered in the millions were described as "commercially extinct" or "dead."

Like a feather on the breeze, fishery populations rose and fell at the

INTRODUCTION

whim of environmental change; yet by the mid-1990s, every downturn was simplistically attributed to a single cause—commercial fishing. All fisheries, not just in U.S. federal waters but around the globe, were "overfished," in "crisis," and anyone who didn't believe that was "in denial." Not to mention, there were "too many boats chasing too many fish."

The timing could not possibly have been more optimal for a sportfishing "conservation" group to promote a solution to that "problem."

SPORT WARTS

On an individual basis, professional truck drivers no doubt received more traffic tickets than private drivers simply because they spent more time on the road. The same was true of commercial fishermen, who spent more time on the water than sport fishermen, and were therefore more likely—either intentionally or quite inadvertently—to run afoul of some fishery law.

In sum, however, just as the sportsmen's cumulative harvest of choice fish exceeded that of the seafood producers, so did their tally of violations: Enforcement agents routinely cited private anglers for exceeding their daily bag limits, retaining fish that were undersized or oversized, and fishing without the required license.

Even when sport fishermen followed the law to the letter and returned countless fish to the water that didn't conform to size restrictions, thousands of these "regulatory discards" later died as a result of rough handling and injuries associated with removal of the anglers' barbed hooks.

Recreational fishermen disparaged gill nets made with nylon monofilament webbing as excessively efficient weapons of destruction. Anglers in Louisiana had outlawed monofilament nets in the 1970s, yet they did not apply the same restrictions to their own tackle.

With a conservatively estimated 250 yards of the transparent material on each of their reels, Louisiana's roughly 500,000 saltwater sportsmen were equipped with more than 70,000 miles of monofilament line, enough to stretch back and forth across the United States more than 20 times.

The non-biodegradable monofilament that anglers either lost or discarded into the marine environment entangled and killed birds, sea turtles and other sea life. The same creatures, including dolphins, also died slow and painful deaths after swallowing the sportsmen's hook-embedded baits.

The recreational fishery was clearly not without its own faults. But as the CCA slung mud at their foes, the net fishermen rarely reciprocated. Instead, professional harvesters reminded all who would listen that sound science and economics were on their side. In the mid-1990s, however, they didn't have much of an audience.

BYCATCH

"Bycatch" referred to the fish and other marine life that sport and commercial fishermen caught incidentally to their harvest of targeted species. The Marine Fish Conservation Network and its member organizations introduced the term into the national lexicon, as they worked to reform federal fishery management, and anti-net forces made the most of the opportunity.

If the buzzword was new to sportsmen, the concept was not: After GCCA won them exclusive excess to redfish and trout, the anglers were reminded that some of "their" fish were still being taken incidentally in the nets that commercial fishermen set for other less controversial species.

INTRODUCTION

There was some truth to the claim. But just as sportsmen countered threats to their Second Amendment rights with the aphorism, "Guns don't kill people, people kill people," the same was true of nets.

In the right hands, in the right circumstances, gill nets could be deployed in such a manner that they produced purer catches than could any other fishery, including that of the hook-and-line anglers. When, for instance, Louisiana's netters ran their large-meshed gill nets around churning schools of mature black drum, absolutely nothing but mature black drum came back in those nets, and they were all of a uniform size.

When strike netters wrapped up tight schools of mullet, on their way to the Gulf to spawn, they might snag a trout that lingered beneath. Was that catch meaningful, in light of the millions taken by sport fishermen?

Scientifically minded fishery managers generally answered "no." For the anglers who regarded every fish taken by commercial fishermen as a diminishment of their own opportunity, if not their very manhood, it was hard to be so objective. Prodding by outside interests didn't make it any easier.

Although GCCA billed itself as a "grassroots" organization, and was even honored as one of the best—by the American Sportfishing Association—the group in fact operated from the top down, not from the bottom up. But its members hardly cared that the corporations were pulling the strings—when they received their orders to "Go get 'em, boys," they were only too happy to comply.

While pictures of gamefish in nets incited anglers, the lurid images of even more emotionally appealing creatures—entangled in gill nets—turned off the very public that the fishermen needed for support.

In Louisiana, the bycatch of endangered sea turtles in finfishing nets was so widely regarded as insignificant that not even the GCCA tried to portray it as otherwise. However, in Florida, where the group's affiliate, the Florida Conservation Association, was trying to convince the public to vote out both fishing nets and shrimp trawls, graphic images of turtles in nets were the centerpiece of its campaign.

Of the many net fisheries in Florida, there was in fact one on the Atlantic Coast where pompano fishermen soaked their gear, mostly at night, in proximity to the turtles' nesting grounds. They could sometimes entangle a sea turtle though by no means did such an "interaction" always result in a fatality.

When Florida's well-intentioned voters banned nets, that fishery was eliminated along with the others.

Alternatively, a similar net fishery on North Carolina's coast, as well as the Gulf's shrimp fishery—where trawlers sometimes scooped up turtles in the tube-shaped nets they dragged behind their boats—provided an example of how professional fishery managers, employed by the public, dealt with such issues:

To accurately measure the extent of the problem they placed observers aboard the fishing vessels. Then they mandated gear modifications, seasonal and areal closures to minimize it, "to the extent practicable." The public thus continued to benefit—economically and nutritionally—from the fishery, while the turtle population was protected, from netting, at least.

Of all possible bycatch, that of marine mammals elicited the strongest emotions from the public, which had zero tolerance for such occurrences. Professional fishery managers, again, sought compromise

INTRODUCTION

as opposed to outright bans, even in fisheries where the incidental take of charismatic marine mammals was nearly unavoidable.

In the eastern Pacific Ocean, purse seiners set their nets around conspicuous schools of dolphins, to catch the tuna that swam beneath. Some of those dolphins died before they could be released, and in the late 1980s and early 1990s, environmentalists campaigned against the practice. Seiners eventually modified their gear and fishing techniques to minimize the mortality of dolphins, which allowed them to continue fishing and market their catch as "dolphin safe."

In the course of those tuna wars, anti-net forces came to the realization that the American public—which in fact annually killed millions of creatures, from ants to antelope, incidental to its driving—would not tolerate the death of a single smiling porpoise in the nets of a commercial fisherman. So their public education campaigns made liberal use of images depicting just that.

For those with a longer view of resource management, the bycatch issue ultimately boiled down to a single question: Was imperfection sustainable?

SUSTAINABILITY

The Worldwatch Institute's report, "State of the World 1990," described a sustainable society as one that satisfied its needs without jeopardizing the prospects of future generations: "Inherent in this definition is the responsibility of each generation to ensure that the next one inherits an undiminished natural and economic endowment."

Amid all the propaganda and crisis mongering of the 1990s, it was easy to lose sight of the fact that the fishery resources that were being fought over in Louisiana were still abundant, after centuries of commercial fishing.

LET THE GOOD TIMES ROLL

Clearly, the fishermen in Louisiana's coastal communities had long "satisfied their needs" without either jeopardizing "the prospects of future generations" or diminishing their "natural economic endowment."

Fault-finding aside, their way of life was, by definition, sustainable, and had been so since they'd first established their communities on South Louisiana's shores in the late 18th century.

The challenge was to continue that sustainable way of life in the face of changes that had occurred since then. At the close of the 20th century, the most threatening changes were political and economic in nature. Longer term, the fishermen's gravest threats were the same environmental changes that threatened the populations of the renewable fishery resources they depended upon.

THE HOLY TRINITY

Fish populations were limited by their natural and fishing-related mortality, the quality of their water, and the quality and area of the specialized habitats that each species required.

Of that Holy Trinity, fishing mortality garnered by far the most attention: Not only did it most directly impact the population but it was the factor over which managers had the most control.

Ironically, fishing mortality—over the long term—was the least significant of the three factors: If regulations were too lax and a species became "overfished"—which meant in essence that it wasn't producing as many pounds as it might—managers could tighten restrictions and leave more individuals in the water to reproduce. Fish, which were typically quite prolific, usually bounced back quickly in response to such measures. Keeping the waters clean and the habitat expansive was seldom so immediately gratifying.

INTRODUCTION

Gulf Coast commercial fishermen were typically *laissez faire* when it came to the practices of other industries though some did individually protest against polluting activities. But simply by pulling food from the water and selling it, everyone in the industry deterred pollution in that the enhanced value of the fisheries made it more onerous for those who contaminated them.

That said, fish appeared to be amazingly resistant to pollution, and often continued to reproduce even after toxins rendered them unfit for human consumption. Habitat was more limiting.

Habitat was the specialized environment—other than the water—that fish required to complete their life cycles. On the Gulf Coast, that included oyster reefs, submerged beds of aquatic vegetation and most important of all, wetlands.

The greatest threats to fishery habitat, especially wetlands, were almost always associated—either directly or indirectly—with development. Louisiana's coastal marshes, for instance, began to disappear after the Mississippi River was lined with levees which were built to protect the people's own habitat from flooding.

As with pollution, the commercial industry, by its very existence, impeded rampant development within coastal communities by exporting seafood and bringing in only money. Removing that impediment opened the floodgates.

As the population grew, so did the pressure on coastal habitats and water quality. Even in locations where the commercial industry was still strong, neither its *de facto* resistance nor the advocacy of individual fishermen, were sufficient to stem the slide.

More comprehensive, sustainably funded efforts were clearly required, and by the dawn of the 21st century it was obvious that the logical source for at least some of that funding was an industry that was by definition sustainable.

The trouble was, a popular movement was emerging at the same time and its members envisioned a future without that industry.

HERE WE GO

In the early 1990s, the Florida Conservation Association demanded that the Sunshine State's fishery managers ban most commercial fishing nets. Both the Marine Fisheries Commission and the Florida Legislature refused on the grounds that there was no scientific basis to do so. The FCA's leaders then gathered enough signatures to put the issue on the ballot and took their case directly to the people.

A prime mover behind the initiative owned the slick and glossy *Florida Sportsman* magazine, one of the most widely circulated publications in the state, and an invaluable platform for "educating" sport fishermen and the general public about commercial fishing.

In November, 1994, Florida's voters overwhelmingly approved the FCA's constitutional amendment, which banned most finfishing nets and shrimp trawls from the state's coastal waters.

The FCA had downplayed the impact of its proposed ban on the commercial industry, informing voters that only a few hundred of the roughly 5,000 net licenses issued by the state of Florida were held by truly professional fishermen. After the referendum, however, the Coastal Conservation Association began to play up the threat of an invasion by "5,000" of Florida's displaced fishermen into the waters of neighboring Alabama, Mississippi and Louisiana. To head them off, leaders of the Houston-based group proposed that those states all ban nets as well.

INTRODUCTION

The three central-Gulf states therefore grappled simultaneously with the issue in 1995. While the CCA tried to stampede their respective management bodies into banning nets, the biologists in each of those states' fishery agencies steadfastly maintained that there was no scientific rationale to do so.

In "Sportsman's Paradise," that wouldn't prove much of a deterrent.

• *Part One* •

LET THE GOOD TIMES ROLL

It has been a grave disappointment to me that in Louisiana, with its fertile waters, abundant natural resources, and its reputation for friendly people, that such vicious rhetoric has been prevalent throughout the debate on this subject. I sincerely hope that some reasonable resolution can be found for this squabble over who can harvest it or how it is harvested, so that the public can join with us in addressing the more important issue of the health of the resources. The Department has an abundance of scientific information regarding our resources, and perhaps if the issue were not so shrouded in rhetoric and emotion, the true state of these resources would be more evident to the general public.
—Joe L. Herring, Secretary,
Louisiana Department of Wildlife and Fisheries, 1995

Our fisheries resources are currently being over-harvested by gill-netters to the point that this vital natural resource is on the brink of disaster.
—Jeff Angers, executive director,
Gulf Coast Conservation Association, 1995

For us, it's very simple: If the decisions are based on science and biology, we win. But if we resort to back-room shenanigans, where politics rules the situation, then we could lose again here in Louisiana. We could go back 50 years into the dark.
—Karl Turner, executive director,
Louisiana Seafood Promotion & Marketing Board, 1995

CHAPTER ONE

SEAFOOD FACTORY

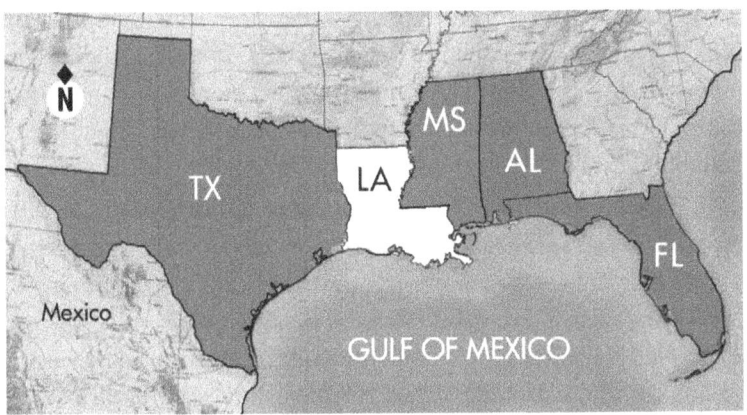

Some years after the collapse of the Soviet Union, the California-based Center for Citizen Initiatives partnered with the Slidell Rotary Club to organize a tour for a group of Russian entrepreneurs; they hoped to show their guests how to create new production, and how democracy worked in the land of the free. So the Rotarians brought them into, of all places, the Louisiana State Legislature.

It was the spring of 1995 and the curious Russians arrived just in time to hear John Hainkel, the senior senator from New Orleans, exhorting his colleagues to "Ban the Nets!" As for the repeated declarations by the state's professional fishery managers that there was no rational reason to do so, the senator asked his peers, "Since when has knowledge been a requirement for what we do up here?"

LET THE GOOD TIMES ROLL

In Alabama and Mississippi, resolving the netting controversy fell to specialized panels with appointed members who were at least partially buffered from political pressure. But in Louisiana, the elected politicians themselves would decide who'd get the fish. As frightening as that prospect seemed to commercial fishermen, the recent referendum in Florida had demonstrated that it could have been worse.

"It puts us in a better position, competitively, than Florida, because there, the other side had to convince the people of the state that they were voting in their best interest," Louisiana Seafood Promotion and Marketing Board executive director Karl Turner told *National Fisherman* magazine. "That kind of a campaign is extremely expensive to put on, and extremely expensive for commercial fishermen to counter.

"So we're in a better position in that respect, because we have 150 legislators, and we're not having to put together a media campaign to reach all four million people in the state of Louisiana."

However, all of the legislators would be up for re-election a few months after the 1995 session, which didn't help the fishermen.

"Y'all have science and biology on your side," Sam Theriot, chairman of the House Natural Resources Committee, admonished net fishermen. "But the other side—the 'sports'—have money and numbers, and that's what I need to get elected."

"There's plenty of reason to be scared," a biologist with the state's fishery management agency told the fishermen, before the session opened. "I've talked to a number of legislators, and they know what's right. Well, they sense what's right. But they feel that they can't do what's right because of all the pressure on 'em."

Whether it was a direct request from his district's Boss Hogg—who'd financed his campaign and thereafter kept him on virtual

SEAFOOD FACTORY

retainer—or a bombardment of beseeching, even threatening, phone calls from the Gulf Coast Conservation Association's lower echelon, the average legislator was under far more pressure to appease his local constituents in the short term than he was to take the high road in the long. And that's assuming he was personally neutral on the issue and needed any convincing.

In "Sportsman's Paradise" many of the state's legislators, including most members of the key House and Senate Natural Resource committees, were anglers themselves some of whom enjoyed being taken fishing by the GCCA and the group's partners, the charter-boat captains.

After the 1994 Annual GCCA Legislators' Invitational Fishing Rodeo, House Natural Resources Committee chairman Theriot gushed in the group's newsletter, "The best Legislators' Fishing Rodeo yet! The fishing was great, but the fellowship was superior. Thanks for everything!"

To keep their nets in the water, Louisiana's fishermen had only to get an acceptable "compromise" bill past these folks. And then the fate of one of the nation's most prolific fisheries would fall to flamboyant gambling Governor Edwin W. Edwards.

Louisiana's commercial harvest in 1994 was second in volume only to that of Alaska. While those landings included fish caught from inland fresh waters as well as federal waters offshore, the lion's share was harvested from the state's fertile coastal zone.

Spanning about 400 miles, from the state of Mississippi to the Texas line, Louisiana's "coast" was less a sharp beach line than a shifting and soggy transition from the sea to solid ground. Unlikely to ever become a Spring Break Mecca, funky coastal Louisiana, with its mud and mosquitoes, marsh grass and murky, olive-colored water, was a seafood factory.

Of the 1.7 billion pounds of fishery products landed in 1994, the herring-like menhaden comprised 1.5 billion; before processing, that harvest was valued at $73 million. Most of these "pogies" were caught with purse seines—high volume and labor-intensive nets that were operated by corporate outfits like Omega Protein and Daybrook Fisheries. The companies rendered the low-valued fish into meal and oil, which were used in a variety of industrial products ranging from plastics, paints and resins to livestock, poultry and fish feed, and nutritional supplements rich in omega-3 fatty acids.

In contrast to Florida where finfish had been the most valuable food fishery, shrimp was number one in Louisiana. In 1994, nearly 13,000 commercial shrimpers hauled in more than 90 million pounds of the succulent crustaceans, which brought them $160 million at the dock. Over 3,000 commercial crabbers trapped 36.6 million pounds of blue crabs, worth $22.5 million, and about 1,900 oystermen produced over 11 million pounds of "meats"—oysters minus their weighty shells—which paid them $20.2 million.

Combined, the menhaden, shrimp, crab and oyster fisheries accounted for about 1.64 billion pounds of 1994's total 1.7-billion-pound haul. The catch of the popular freshwater crayfish added 24 million pounds—worth $30 million—while the remaining 38 million pounds was mostly finfish.

[Facing page] Menhaden seiners at work. Louisiana's largest commercial fishery was based on the wetland-dependent menhaden. Used for industrial purposes, the small, oily fish was valued at just pennies per pound. The sportsmen's proposed net ban exempted this corporate fishery and targeted the family fishermen who netted more valuable edible species. *(Brian Gauvin)*

SEAFOOD FACTORY

LET THE GOOD TIMES ROLL

About 40 percent of the state's finfish harvest was comprised of freshwater species and saltwater fish from offshore. Major freshwater fisheries included catfish—4.4 million pounds worth $2 million—and major offshore fisheries included hook-and-line-caught yellowfin tuna—3.8 million pounds worth $10 million—and red snapper—1.7 million pounds at $3.4 million.

The remaining 60 percent of the finfish haul came from Louisiana's nearshore coastal waters: It included a complex of about 20 species which totaled more than 23 million pounds in weight, was worth nearly $15 million, and was harvested almost exclusively with nets.

FISHERIES AT RISK: OLD AND NEW

The state's coastal finfishermen had traditionally targeted a handful of culinary favorites like redfish and spotted seatrout, pompano and flounder. But as sport fishermen encroached on their bread-and-butter fish in the late 1980s, commercial fishermen began to shift their efforts to other of the many abundant and lesser appreciated species that inhabited, or migrated through, the state's coastal waters. Assuming that harvesters received reasonable allocations of their traditional species, and could continue to develop these underutilized fisheries, Louisiana's coastal finfishing industry had plenty of room to grow.

Markets for at least nine coastal species had been well established by 1994, when their combined landings exceeded 22.7 million pounds, worth more than $14.3 million at the dock. At 12.6 million pounds, the mullet fishery was the largest, followed by black drum (3.74 million pounds), Atlantic sheepshead (3.3 million pounds), spotted seatrout (1.02 million pounds), flounder (975,000 pounds, with most taken incidentally in shrimp trawls), alligator gar (569,000, including

SEAFOOD FACTORY

many from inland fresh waters), king whiting or "ground mullet" (460,000 pounds, with many taken incidentally in shrimp trawls), Florida pompano (115,000), and red drum, which at the time was under a commercial moratorium.

Room for growth among these traditional offerings may have been more limited than for the many species that were still comparatively neglected, yet it was hardly insignificant.

For example, when the state interrupted the commercial red drum fishery in 1988, net fishermen shifted their efforts to the closely related black drum, to satisfy demand for "blackened redfish." In response, the Louisiana Department of Wildlife and Fisheries developed its first comprehensive management plan for the species in 1989: it allowed netters to bring in 3.25 million pounds of the more desirable "puppy" drum—under 27 inches long—and another 300,000 head of the larger "bull" black drum. Averaging 15 pounds apiece, those bulls would weigh a total 4.5 million pounds, bringing the potential sustainable harvest to 7.75 million pounds, about 4 million more than was reported in 1994.

State biologists also considered the striped mullet resource to be underexploited in most areas of the state, suggesting that its harvest could be substantially increased over that of 1994. Also, in 1994, state biologists declared that the red drum population could safely accommodate a commercial harvest of 3.2 million pounds, a considerable increase over the zero pounds that the Legislature then allowed.

At least 11 far less controversial species offered even greater growth opportunities, as suggested by their 1994 landings, which totaled 468,000 pounds, and earned commercial fishermen just over $258,000.

In descending order, by volume, those species included sand or "white" seatrout (236,000 pounds), Spanish mackerel (83,000 pounds), Atlantic croaker (74,000 pounds), spot (32,000), blue runner (22,000), bluefish (11,000), sea catfish (4,800), rays (4,600), pinfish (500), crevalle jack (40) and ladyfish (0).

Some of these fish, such as the sea catfish, which included both the gafftopsail and the hardhead, were considered downright objectionable to sportsmen, who likely would not have opposed a thousand-fold increase over 1994's landings of 4,800 pounds. Landings of various rays and the underutilized blue runner, bluefish and crevalle jack could likewise have been multiplied, and the migratory ladyfish, which Alabama and Florida fishermen were producing for export to Asia, could have easily withstood a Louisiana harvest of at least a million pounds, as opposed to the zero reported in 1994.

Even the Spanish mackerel, which was marginally popular with Louisiana's saltwater anglers, had great possibilities: The federal National Marine Fisheries Service set an annual gulf-wide Total Allowable Catch for the migratory species. Of the 9-million-pound TAC, 5.2 million pounds were allocated to the commercial fishery and 3.8 million to the sportsmen. Yet the combined haul by both sectors averaged less than 5 million pounds, which left more than 4 million pounds that could safely be harvested. In 1994, mackerel landings in Alabama and Florida's west coast—the Gulf's two biggest producers—totaled roughly a quarter-million and 2.4 million pounds, respectively, which left about 2.5 million pounds of the annual commercial allocation uncaught.

Over time, Louisiana's net fishermen likely could have more than doubled their 1994 catch to a conservative and sustainable 50 million pounds—including roughly 25 million of traditional and 25 million

SEAFOOD FACTORY

of underutilized species—while the sportsmen continued to do their thing. That assumed, however, that the productivity of the state's coastal waters didn't decline.

VANISHING WETLANDS

Geologically speaking, coastal Louisiana was brand new. Totally alluvial, the land was built of sediments carried by the Mississippi River from as far north as Canada. As one delta grew too large, the river abruptly changed course to seek a shorter route to the Gulf, and began to build another.

The Teche Delta of southwest Louisiana began to build about 6,000 years ago; the St. Bernard Delta, northeast of the river's present location, was the second; the Lafourche Delta, to the west of the present river was about 3,500 years old; and the modern delta, along the course of today's Mississippi River, was the newest of all.

Owing to society's intolerance of flooding, the modern delta was no longer building, and most of the older deltas were rapidly vanishing.

After America's Great Flood of 1927, the federal government built high earthen levees along the river, and they worked—no more flooding. But instead of depositing its nourishing sediments over the lowlands, the river ejected them into the sea. Instead of building, South Louisiana began to shrink.

Since the soils in the recently formed deltas were continually compacting, the land naturally subsided. Before the levees, the annual spreading of sediments more than compensated for the sinking. After, the land could only go down.

Then, in the 1930s, the discovery of oil triggered a succession of additional insults that exponentially accelerated the land loss.

The oil companies pumped deposits of petroleum, brine and natural gas from beneath the land, which hastened the sinking. To gain access to the well sites and to run pipelines, they dredged canals through the marsh, heaping the mud along both edges in high banks. The weight of those spoil banks in turn shut off the tidal flow through the soils beneath, which killed the grasses that held the marsh together. As the resulting expanses of dead marsh reverted to open water, winds churned up the shallow bays, and the sediments hemorrhaged from the wetlands through the deep and unnaturally straight canals.

As the marshland deteriorated, it became more vulnerable to storms with waves that simply rolled up the remaining grass patches.

The Louisiana coast shrank by more than 30 square miles per year just between 1956 and 1978. Although the rate of land loss subsequently slowed—to an estimated 16.75 square miles per year—it still equated to the area of one football field per hour.

Confronted with the prospect of losing the southern portion of the state, a diverse coalition of organizations, in 1989, cajoled the government into establishing a trust fund to restore Louisiana's coast. Logically, the centerpiece of the program was to reintroduce the river's sediments into the system. But the freshwater diversions were massive projects that involved several governmental agencies and typically took decades to complete.

Frustrated with the slow process, and fearing for both their livelihoods and homes, commercial fishermen in one of the state's more progressive trade organizations, in late 1994, began to advocate for a more streamlined and localized action plan to help retard the land loss, until the massive projects came on line. While the Organization

SEAFOOD FACTORY

of Louisiana Fishermen anticipated future financing from a diversity of interests, including the sport-fishing and oil industries, the group's initial proposal tapped only the commercial sector, through either the purchase of habitat stamps or the collection of severance taxes on seafood products. The funds were to hire local contractors for projects that were to be developed by the group's proposed habitat restoration panel within the Department of Wildlife and Fisheries.

The timing for their sustainable-use proposal couldn't have been better because Louisiana's coastal fisheries in the mid-1990s may never have been more abundant.

Historically, the Mississippi's fresh water flowed continuously

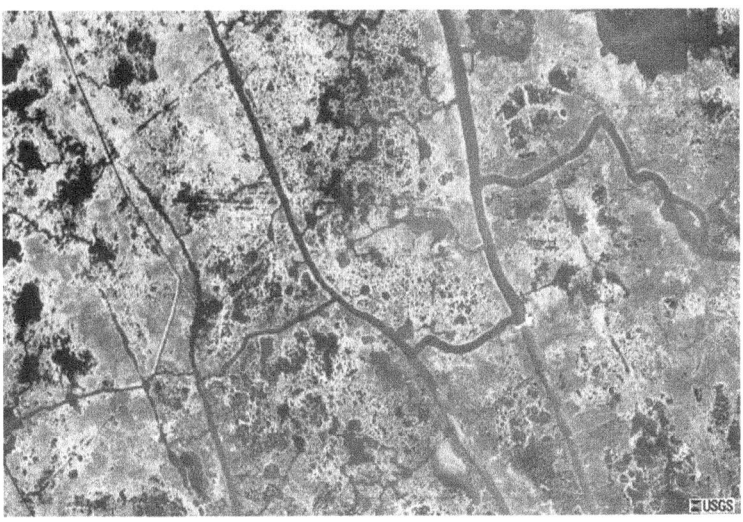

From nearly solid ground, interspersed with well-defined bayous, Louisiana's coastal marshes were deteriorating into a lacework of ponds, as illustrated in this 1990 photo. *(Microsoft Terraserver Images & U.S. Geological Survey)*

through the coastal marshes, which limited the range of saltwater seafood species to the very edge of the Gulf.

When the levees were built along the river, its tributaries were sealed off. No longer held back by the steady flow of fresh water through the system, the Gulf's salty waters began to intrude, and the nursery area for the state's marine seafood species expanded.

In addition to the saltwater intrusion, the ongoing subsidence created even more fishery habitat as depressions in the marshland sank into potholes, ponds and bays. From a virtual prairie, etched by a few well-defined bayous, the marsh took on the appearance of lace, and every new hole, edged with grass, represented additional living space.

Wildlife species often benefit when man alters the environment. The trick is to sustain that benefit. In this case, thanks to man's intervention, the populations of wetland-dependent seafood species ballooned through the entire twentieth century. But as the grasslands continued to crumble into the Gulf, a day of reckoning loomed.

The conversion of inland freshwater swamps to saltwater marsh postponed that day but unless the problem was addressed in a serious and ongoing fashion, Louisiana's fishery productivity faced inevitable decline.

Meanwhile, in the early 1990s, after generations of fishermen netting with trammel nets, gill nets, and seines, Louisiana's coastal fish were thick. Of the 20 or so species that were in play, only the redfish was considered to be overfished; though it had been exclusively a recreational species since 1988, commercial fishermen still had a shot at getting back their wintertime staple, which was producing record crops. The only other species that may have been close to being "fully utilized" was the sportsmen's favorite, the spotted seatrout.

SEAFOOD FACTORY

Netters and consumers were allocated a million pounds of this culinary favorite, and the only other species under a quota was the black drum. While fishery managers developed sustainable management plans for the state's other major fisheries, they were virtually wide open.

With a marsh half the size of Massachusetts, within one of the nation's least populated states, coastal Louisiana was a century behind Florida in terms of tourism and real estate development. The few upscale recreational developments stood out like space colonies amid the many humble and isolated fishing communities where fishermen "lived the fish," and their kids dreamed of the day when they'd have their own boats.

Then it began.

CHAPTER TWO

RALLYING THE TROOPS

"Ban Gillnets in Louisiana Waters? If you say so!" began GCCA's September 1993 newsletter to its membership:

> "Recreational fishermen generally agree: indiscriminate gillnetting is destroying our fisheries resources. Many sportsfishermen, therefore, want to seek to ban gillnets from Louisiana waters.
> "The GCCA State Board of Directors will be meeting this fall to discuss eliminating the use of gillnets in statewide waters. This legislative proposal may be voted on as early as the 1994 Regular Legislative Session.... The State of Florida will soon be voting on a constitutional amendment to ban gillnets in Florida waters. If they are successful, Louisiana and Mississippi will be the only Gulf States allowing gillnets.
> "Many gillnetters fishing Louisiana waters are already from Florida. What will happen if Florida conservationists are successful there? As soon as next fall, Louisiana could feel the brunt of a Florida Success."

As GCCA began to fire up the sportsmen, seafood advocates warned the public what was coming. The Louisiana Seafood Promotion & Marketing Board was a division of the state's

LET THE GOOD TIMES ROLL

Department of Wildlife & Fisheries, so its executive director was prohibited from lobbying. Yet Karl Turner did as much as he could to defend the industry, and in his lengthy letter to the editors of the state's major newspapers, he alerted consumers that "There is a national movement that threatens to eliminate all production of domestic seafood under the guise of conservation or 'highest and best use' of limited resources.

"....A new, highly vocal and politically connected group is again engaging in alarmist tactics aimed at closing commercial fishing. They have started the crusade to ban the nets....Consider the national trend: By amending the state constitution, California recently outlawed nets even though the state legislative and fish management agency saw no need for a ban.
"The closure of California's commercial net fishery was due to allegations by a tireless group who hide under the cloak of conservation and claimed 'bycatch and endangered species interactions problems.' It is no surprise that California has the highest number of recreational fishing licenses and boat registrations in the nation.
"...This capricious style of fishery management is spreading. Florida now faces a similar constitutional amendment due to efforts of recreational fishing organizations, outdoor press, and tackle manufacturers. These anti-commercial-fishing-interest-cloaked-in-conservation-garments aim to eliminate all gill and trammel nets, beach seines, purse seines, finfish and shrimp trawls from most Florida waters. Not surprisingly, only those nets used to support recreational fishing will be allowed.
"Fish are a common property resource; they belong to everyone.
"Beware, however, if you are a seafood consumer but don't own a boat or have the leisure time to go fishing. These so-

RALLYING THE TROOPS

called ban-the-net conservationists want the public to believe that it is far more noble that a fish dies on a hook than in a net."

After Turner's letter, a supportive follow-up appeared in the state's major newspapers. William Chauvin, managing director of the American Shrimp Processors Association, and former chairman of the federal Gulf of Mexico Fishery Management Council, wrote:

"I have been very aware of the recreational interests'

Redfish, "speckled trout," and a variety of other wild-caught coastal fish had traditionally been available to consumers in New Orleans via restaurants and retail markets, such as this one photographed in the early 1990s. Trout and redfish "po-boys" — sandwiches made on French bread — were a ubiquitous staple in the city until GCCA rallied sportsmen to monopolize the species.

determination to gain control of the fisheries, and of their self-serving goals....Mr. Turner's letter was primarily directed to alerting the consumer, those Americans who enjoy eating seafood, this healthy bounty from the sea. His message was that the consumer is threatened by certain selfish groups that could in the future restrict seafood to imports, and the consumer would be unable to enjoy the fresh product of our American waters."

These early warnings attracted a hail of responses from sportsmen.

The Coastal Conservation Association served as the national umbrella over its regional associations, and Cornel Arceneaux, of Baton Rouge-based Cornel Arceneaux Insurance, was a CCA trustee. He was also president of the Louisiana Association of Coastal Anglers. Formed in late 1992 to ride the net-ban wave, LACA was a home-grown group of sportsmen, boat dealers, real estate developers, and guides whose militancy made even the GCCA seem reasonable.

Claiming a membership of 1,100 in 1994, the "Lackies," as fishermen naturally dubbed the group, made no pretext of assuming a stewardship role for the state's coastal resources. The single-purposed Louisiana Association of Coastal Anglers was "dedicated to the complete elimination of gill nets in coastal areas," according to a 1994 interview with Arceneaux by staff of the publicly funded Louisiana Sea Grant College Program. The group hoped to attain its goal, he said, by "educating the general public and the Legislature."

In addition to letters in mainstream papers, LACA "educated" by peppering the sporting press with full-paged illustrated ads with alarmist captions like "Plastic Walls of Death versus Redfish and Speckled Trout!" and the sportsmen's classic, "They Might As Well Be Using Dynamite!"

RALLYING THE TROOPS

Louisiana Association of Coastal Anglers advertisement in 1994 *Louisiana Sportsman* magazine both decried the use of nets by commercial fishermen and invited recreational fishermen to participate in the group's fishing rodeo.

LET THE GOOD TIMES ROLL

Arceneaux's mid-August 1993 letter, titled "Gill net culprit in fisheries destruction," appeared in newspapers across the state. He said, in part:

"This is in response to the letters from Mr. Turner (July 31) and Mr. Chauvin (Aug. 8), which berated the conservationists' efforts to protect and preserve Louisiana's fisheries for the future—everybody's future!

"There is a national movement that threatens to eliminate all seafood: the use of gill nets by commercial fishermen.

"Since commercial fishermen began using the monofilament gill net in the 1970s, there has been a devastating reduction in the world's fish population. Some species have been fished into extinction, while many others are overfished to a dangerously low level that threatens their very existence.

"Gill nets are non-discriminating tools that not only kill a wide variety of market species by the thousands but also kill millions of pounds of undersized, threatened or protected fish and other sealife as bycatch. Conservation of fisheries and management of marine resources to enable the species to replenish and maintain their populations is not compatible with the use of gill nets.

"…While nets have been used in the past to harvest fish, they were not the same as what is being used today. Today's use of larger, nearly invisible monofilament gill nets, bigger and faster boats and high technology leads to wanton destruction of our fish populations.

"There is little difference in the end result between the use of gill nets and the use of dynamite or electric shock. States such as Texas, California and Georgia as well as most of the Great Lakes states have deemed it necessary to ban gill nets. Florida citizens are working to make gill nets illegal, and many Louisiana citizens want Louisiana to do the same.

"Our fisheries are not inexhaustible resources. Like the

RALLYING THE TROOPS

buffalo and the wild turkey of the past, our fisheries are being reduced to near extinction.

"Decommercialization of species has been an acceptable method of conservation for many species and is now being used for redfish. The species is making a comeback. If this method of control is to be avoided, the commercial fishermen need to fish in a way that is responsive to the needs of the dwindling fish stocks.

"...The seafood consumer would benefit from any changes in the seafood industry that would increase the size of local fish stocks.

"...Nature's balance is being destroyed by man's inventions and greed. Serious public action is necessary to stop the plundering of the world's fish populations. Banning gill nets is the most important first step to solving the problem."

In the same vein, fellow LACA officer Beau Weber, an insurance salesman from suburban New Orleans, followed up with a mid-September letter titled, "Preserving our marine resources for the future." It included:

> "The worldwide seafood industry is in dire straits due to years of overfishing and improper regulations, and a lot of commercial fishing industries have put themselves out of business through overfishing. The term that describes this condition is commercial extinction (when a resource is depleted to the extent that it can no longer support the industry).
>
> "Is it wrong to attempt to protect a resource before there is nothing left of it, to deplete a resource to an extent that it cannot replenish itself and therefore becomes extinct? Is it wrong to try to protect a species from extinction?
>
> "Conservation groups in Louisiana have attempted to

protect redfish and speckled trout from commercial extinction due to overfishing. For this, they have been condemned by commercial fishermen who want the right to catch every fish in the water in order to satisfy their greed.

"…Commercial fishermen should be allowed to catch and sell speckled trout and redfish, but they should do it the way everyone else does—with a rod and a reel. Eliminate the lethal weapons of the fishing industry. Ban the gill nets and preserve fishing for our children."

CHAPTER THREE

REDFISH: BONE OF CONTENTION

By the fall of 1993, more than a year before the Florida vote, the sportsmen's leaders were planning to parlay their anticipated "Florida Success" into a "Louisiana Success," and tried to rally the Bayou State's anglers. If that was a little premature for most sportsmen to get excited, when state biologists announced, later that winter, that commercial fishermen could once again net some redfish, the sportsmen quickly snapped to attention.

Reds had been off limits to seafood producers since 1988, when the state—after federal prompting—closed its entire red drum fishery while managers developed rules to relieve both sport and commercial pressure on the species. Though the recreational fishery was shut down for just four months—from early February to June 1, 1988—sport fishermen were enraged to have been punished for what they perceived to be the sins of the commercial sector. In retaliation, GCCA-led anglers rammed a "sunset gamefish" provision through the 1988 Legislature, which allocated the species exclusively to sportsmen "for three years."

In 1991, when the "temporary" closure was scheduled to sunset, commercial fishermen were hardly surprised when GCCA claimed that there weren't yet enough reds to share. Then biologists with both

LSU and the Louisiana Department of Wildlife and Fisheries essentially agreed, stating that the overall harvest couldn't yet be safely increased, and that any allotment to the commercial sector would have to come out of the recreational sector's share.

Like most agency biologists east of Texas, Louisiana's biologists had always maintained their neutrality on the question of allocation: "A dead fish is a dead fish" was their mantra, and they hadn't abandoned that stance in 1991. So when they said, "Better wait," *everyone* listened, and an untimely fish kill associated with a December 1989 freeze had made it even harder to argue with them.

Rather than wait, or honor the 1988 "sunset" agreement, Senators John Hainkel, of New Orleans, and Larry Bankston, of Baton Rouge—both GCCA supporters—pushed a bill to designate reds as gamefish *permanently*. But the commercial industry was still strong enough in 1991 to amend GCCA's bill to say that reds would remain gamefish only until biologists said the harvest could safely be increased.

Eliminating any ambiguity, the bill directed the state's biologists to deliver an annual report to both the Wildlife and Fisheries Commission and the Legislature, which was to include a stock assessment of the redfish and "a total allowable catch with probable allocation scenarios."

In 1992, the biologists said, Not yet. And in '93: Nope.

Then, in February of '94, they announced to the Wildlife and Fisheries Commission that, if recreational fishermen were held to their daily bag limit of five redfish, commercial fishermen could safely harvest up to 3.2 million pounds while maintaining the 30 percent minimum escapement rate recommended by the federal government.

The federal National Marine Fisheries Service had become

REDFISH: BONE OF CONTENTION

involved in the mid- to late-1980s, after data obtained from the new offshore purse-seine fishery indicated that the number of young fish escaping from state waters, across the Gulf, was insufficient to sustain the spawning stocks found in federal waters. Federal biologists recommended that the states tailor their regulations to allow at least 30 percent of their succeeding crops of redfish to survive long enough to migrate offshore. By 1994, escapement from Louisiana's waters exceeded that level to the extent that state biologists, at the February meeting, told the commission that a conservative 50 percent rate could still be maintained while allowing the commercial sector to harvest 1.6 million pounds.

The state's red drum fishery had historically operated under few restrictions because they weren't needed. But as the number of sport and commercial fishermen grew over the years, a regulatory re-set became necessary. Thanks to politics in Sportsman's Paradise, however, the new restrictions weren't fairly applied.

Commercial landing records had been kept since 1887, when the industry reported a catch of 289,000 pounds. Landings slowly increased though the century, and surpassed one million pounds for the first time in 1973. The commercial catch then averaged about 1.5 million pounds each year until a convergence of events in the 1980s caused landings in the state to balloon, then dry up.

Discord among the Arab oil producers dropped the price of oil from $40 a barrel in 1982 to a low of $10 in 1986. As the petroleum industry bottomed out, an army of unemployed oil workers took to the water to try to pay their bills. Commercial landings reached 2.6 million pounds in 1984 and nearly 3 million pounds in 1985. Then, in 1986, they reached an all-time high of more than 7.8 million pounds, which did include several million pounds that had been purse seined in federal waters and landed in Louisiana. When that federal fishery was shut down, in the midst of the blackened redfish craze, demand increased for fish from the state's waters.

In 1987, a decade after the Texas Legislature had begun to rein in its fishery, Louisiana's Legislature imposed similar restrictions, including a first-ever annual quota of 1.7 million pounds on the commercial fishery; the quota represented the industry's ten-year harvest average, with the highest and lowest years thrown out. The same bill also raised the minimum size limit for commercial fishermen, from 16 to 18 inches. Confusion over when the new quota was to resume—retroactively, on January 1, or on September 1, when the law took effect—resulted in the industry landing over 4.5 million pounds in 1987.

Legislators restricted the recreational fishery at the same time they did the commercial. In 1987, they began to phase in first-ever minimum size limits for anglers: from an initial 14 inches, the minimum increased to 15 inches in 1988, and 16 inches in 1989.

From 1980 through 1986, fishing under a daily bag limit of 50 redfish, the sportsmen's take averaged nearly three-fourths of the total harvest; before the minimum sizes went into effect their redfish were averaging just 7 inches in length.

When the new size limits were imposed, state biologists projected that they would reduce the sportsmen's take by nearly 50 percent, which proved accurate: From more than 1.5 million redfish in 1987, recreational landings fell to nearly 813,000 fish in 1988 and 616,000 in 1989. (The reduction in the total *weight* of the recreational catch was not nearly as pronounced because the new size restrictions quickly led to an increase in the average size of the fish caught: Anglers landed

REDFISH: BONE OF CONTENTION

3.6 million pounds in 1987, 2.5 million pounds in 1988, then nearly 4.4 million pounds in 1989.)

With some tweaking, those initial measures would have maintained the requisite conservation goals, while keeping both commercial and recreational sectors in the fishery. Indeed, Wildlife and Fisheries biologists told the Legislature in 1988 that commercial fishermen could safely harvest 900,000 pounds of redfish. Instead, sportsmen elbowed the commercials out of the fishery altogether: After passage of GCCA's 1988 "sunset gamefish" bill, commercial landings tapered off to 245,365 pounds that year, 24,811 in 1989, and effectively zero after that.

In addition to barring commercial redfishing, the gamefish bill left even more fish in the water than had the 1987 size restrictions: It reduced the bag limit for sportsmen—who resumed fishing on June 1—from 50 to 5 fish, and only one of those could be over 27 inches, which further protected larger redfish.

In response to all these restrictions, the annual escapement rate by 1994 approached 70 percent, more than twice the conservation standard. In the five years from 1989 through 1993, sport fishermen brought in nearly 25 million pounds, an annual average of 5 million pounds, while commercial fishermen harvested virtually no redfish. So they didn't think it unreasonable that they now should be able to bring in a few as well.

Anticipating the biologists' positive report, the commercial fishing industry presented a re-entry plan to the Commission that was sweetened considerably over the plan it had presented in 1991, when they first expected to be dealt back into the fishery.

LOUISIANA'S **RECREATIONAL** HARVEST OF RED DRUM
(Since Commercial Fishermen were Excluded from the Fishery)
1989–1993*

YEAR	POUNDS	NUMBER OF FISH
1989	4,379,774	1,052,081
1990	3,014,515	616,604
1991	3,999,416	872,713
1992	5,833,875	1,767,938
1993	7,424,691	1,913,832
TOTAL:	**24,652,241** POUNDS	**6,223,168** FISH

Commercial fishing in the state ended on January 15, 1989.

As part of their 1991 package, commercial fishermen supported and passed a bill in the Legislature that distinguished their nets from those used in the state's sizeable inland freshwater fishery, and raised the cost of the saltwater net license ten-fold, from $25 to $250 a year. Funds from the fee increase were dedicated to implementing an improved system for reporting their landings.

In 1991, the last year the freshwater and saltwater licenses were lumped together, the department sold a total of 2,816 resident and non-resident gill-net licenses. In 1992, when the law went into effect, the agency issued 907 resident and non-resident "saltwater gill-net" licenses. Managers now had a better handle on how many fishermen

REDFISH: BONE OF CONTENTION

were out there and, in this open-entry fishery, the increased fees culled out at least some of the recreational netters, part-time oilfield workers, and other fly-by-nights who gave the industry a black eye, mostly by illegally leaving set gill nets unattended.

The fees were intended to finance the implementation of a sales card program: When making a sale, commercial fishermen would hand over their plastic card which dockside buyers would swipe with a machine, producing several copies of the transaction, one of which went directly to the LDWF.

Not only would the automated system prove more accurate than the hard-copy, voluntary system then in place, but its real-time reporting was essential for tracking the landings of species, such as redfish, that were to be managed by quota.

The law had stated that the fees were to be collected by Wildlife and Fisheries, kept in a designated account, and used for the acquisition of the sales cards and the machines that seafood buyers would use to record their sales. The money was also to be used for "enforcement" which, to the bill's authors, meant the periodic auditing of the buyers' records. Instead, when the time came to implement the program, fishermen learned that higher-ups in the department's law enforcement division had siphoned off the funds—more than $600,000—for their own operations.

Thwarted in their 1991 effort to fund an improved reporting system with increased net-license fees, the fishermen, in their 1994 plan, tried another approach: To fund "the creation and operation of the Commercial Fisherman's Sales Card reporting program," they proposed that $400 out of each $1,000 wholesale dealer's license be applied to the program, as well as the entire $200 cost of a proposed endorsement that would be required of any fisherman wanting to harvest either redfish or trout.

To be eligible to purchase the endorsement, and to prevent a rush into the fishery, fishermen had to qualify by having held saltwater net licenses in prior years. The annual commercial quotas for trout—or red drum, "should the prohibition be lifted"—were to be divided equally among all the fishermen who obtained endorsements. Netters would receive a tag for each of their fish, and the tags could be sold or leased to other fishermen. Participants would be required to report their trout and redfish landings to the department monthly, or their endorsement would be revoked.

Nationally, individual transferable quotas, or "ITQs," were just coming into vogue as a fishery management tool. They were seen as a means to eliminate the "derby" fishing that occurred when fishermen raced to fill a single industry-wide quota.

But the fishermen's progressive ITQ proposal didn't impress GCCA. Testifying before the Wildlife and Fisheries Commission, the group's executive director Mark Hilzim—formerly in public relations at defense contractor Lockheed Martin—repeated the 1991 assertion that it was still too soon to share.

With the state agency's data no longer in support of that claim, GCCA produced outside testimony to back up its position.

The Texas Parks and Wildlife Commission's Chairman Ygnacio Garza advised a "go-slow approach to any liberalization efforts…In Texas, it's been determined that the social and economic benefits to the State are maximized by making red drum a gamefish. This allocation decision continues to be justified given the increasing recreational fishing demand and the difficulty experienced by law enforcement in controlling fishing mortality in a commercial net fishery."

REDFISH: BONE OF CONTENTION

The University of South Alabama's Bob Shipp, who was vice chairman of the federal Gulf of Mexico Fishery Management Council and a CCA leader in Alabama, where sportsmen had had a lock on the redfish since 1984, offered, "...the implication that this initial evidence of improvement would warrant an immediate increase in harvest is appalling....The stock needs *many* years to strengthen..."

Richard Condrey, an associate professor at LSU's Coastal Fisheries Institute, who also chaired the Gulf of Mexico Fishery Management Council's Red Drum Stock Assessment Panel, testified: "...I urge that we continue to err on the side of the resource where there is an uncertainty which could affect the long-term abundance of the population....While the stocks appear to be recovering, it is the unanimous recommendation of the Red Drum Stock Assessment Panel that we hold a steady course and resist all efforts to reduce the current conservation measures until it is clear that the stocks are in an equilibrium state of recovery."

After the testimony, the three commercial representatives on the seven-member commission voted to implement the industry's transferable quota proposal for re-opening the fishery but they were outvoted by the four sportsmen, who dismissed it as "unenforceable."

Louisiana's Wildlife and Fisheries Commission was an honorary body that lacked the authority to re-open the fishery. It could, however, direct the state agency's staff to develop an Individual Transferable Quota management plan which the fishermen could present to the Legislature.

Undeterred by the proposal's failure to pass, two of the seafood industry's staunchest advocates, Representatives Kenneth Odinet, of Arabi, and Frank Patti, of Belle Chasse—both Democrats who sat on the House Natural Resources Committee—drew up a resolution that directed the Commission to develop an ITQ plan, which they hoped to get passed in the Legislature when it convened in April.

But that plan faced a hurdle as well. Louisiana's general sessions, when any matter could be addressed, alternated with fiscal-only sessions. Since the upcoming 1994 session was for fiscal matters only, a special session needed to be called to deal with the redfish issue. And that required the approval of the governor.

"I could evade this entire question by saying that the upcoming session is a fiscal-matters-only session, but I'm not going to do that," said Governor Edwin Edwards, as he addressed a late-February meeting of the Louisiana Outdoor Writers Association. Instead, he made it quite clear: "I will not call a special session for the purpose of addressing that issue."

Son of a Presbyterian sharecropper and a French-speaking Catholic, the beguiling Edwards personified the two major cultural groups of North and South Louisiana, and was the first and only governor to serve four terms. An attorney from humble origins, the populist Edwards had always been a friend to the fishing industry and, indeed, to most of the state's working people, for which he was rewarded with votes. But with plans to retire from politics after the 1995 elections, votes were no longer his priority.

Parroting GCCA's assertion that there weren't yet enough redfish to share, Edwards told the outdoor writers, whose works targeted primarily hunters and anglers, "From what I understand, we're within a year or two years from a time when it would be prudent to consider a commercial harvest."

The fishermen didn't get a legislative hearing on their redfish resolution, but their push to re-enter the fishery brought the anglers out of the woodwork.

CHAPTER FOUR

CONVINCING THE PEOPLE

A month before the Florida vote, in an early-October article, Bob Marshall, the *Times-Picayune's* outdoor columnist, interviewed LA-GCCA's executive director, Jeff Angers, who'd recently replaced the group's first salaried director. Angers used the opportunity to officially announce that his group was going for a net ban in Louisiana.

"That's our goal for 1995. We're growing by leaps and bounds—our membership has soared to more than 7,500—and the money is rolling in. Basically, the sportsmen of this state are fed up with the nets. They see what's happening in Florida, and they want to protect what we have now, before it's too late. We think we have the organization now, and the strength to get it done....We say rod-and-reel is the way to go. And we think we can get it done here, like it's getting done in Florida."

When the columnist expressed doubt that the Legislature's natural resource committees would abandon their reliance on scientific information, Angers confidently responded, "We're working hard right now on coming up with the science we need to support our position. We think we'll have enough facts to convince the people this is what needs to be done. And if we convince the people, we'll have the votes in Baton Rouge."

LET THE GOOD TIMES ROLL

ORGANIZING A DEFENSE

Unlike commercial fishermen in states such as Florida, North Carolina or Maryland, who'd long maintained statewide professional organizations, Louisiana's were splintered into about a dozen bands across the coast. Each group's self-appointed leaders lobbied legislators directly, but with needs that varied from region to region, the overall result was often a babel of conflicting voices.

Only a universal threat like "redfish" or "nets" brought them together.

To fight the net ban, they formed a coalition that included dockside buyers and urban seafood distributors. After several organizational meetings, the Louisiana Seafood Management Council went public in late October 1994. The meeting in New Orleans attracted about 300 worried fishermen and a couple of their legislative supporters.

Representative Frank Patti reminded the fishermen that GCCA "is well-organized and has all the money it needs" to push the net ban. "You need to organize, and you need a spokesman," advised Patti. "You can't have 40 people talking 40 different ways."

Once they selected their spokesmen, Representative Kenneth Odinet suggested, they should speak to the governor. The recreational fishermen had such a strong lobby in the Legislature that it could take the support of the governor to make a difference, he said. "GCCA does have the numbers, but the government has to realize that people need to eat and they need to support their families."

Odinet further advised the group to begin putting a defense together immediately. "Don't wait until the session begins in March, and then try to piece something together."

New Orleans seafood dealer and industry patron Preston Battistella

CONVINCING THE PEOPLE

GCCA executive director Jeff Angers served for several years as chief of staff for State Representative Louis "Woody" Jenkins, and also worked as an advertising executive at Jenkins' Baton Rouge cable television station WBTR. In 1991, Angers ran for state representative in Baton Rouge's House District 69, which he described in a fundraising letter as "the most Republican seat in the Legislature." Running as a pro-lifer, Angers tried to position his opponent as an "adversary of Louisiana's unborn children." After his defeat, GCCA hired Angers, and he quickly made the transition from saving the unborn to saving the fish. *(Political campaign literature)*

then announced the centerpiece of the council's plan. Since the sportsmen's primary impetus for a ban was the impending invasion of displaced netters from Florida—should voters there approve the Florida Conservation Association's net-ban amendment—an across-the-board moratorium on the sale of new net licenses to *anyone* would thwart that invasion. End of problem.

INVADERS COMING!

The locally owned *Louisiana Sportsman* wasn't as slick as *Florida Sportsman*, the glossy magazine that propagandized for the net ban in that state, but it operated on the same business model by selling advertisements to businesses that marketed goods and services to recreational fishermen. Staff members promoted the ban in their writing, while ads in the tabloid further incited anglers.

In the November issue, a full-paged double spread by GCCA, under the headline, "Abuse of Marine Resources Rampant, Frightening," featured a pair of photographs: In the first, someone, with their face obscured, displayed a single spotted seatrout in a set gill net; in the other, a close-up focused on the fish, hopelessly entangled in what appeared to be brand-new webbing:

"These photos were taken on the beach near Calcasieu Lake during August 1994. The commercial season for speckled trout opens in September. Guess this fish didn't know she was out of season.

"Conflicts between user groups—recreational and commercial—are at an all-time high. Sights like these are a major cause of that tension. When GCCA is successful in its legislative effort to ban gill nets and other entanglement nets, those user conflicts should begin to fade away. Once all fishermen are using the same gear—rod and reel—all will have more equal access to Louisiana's fisheries resources."

On the facing page under "Support the Gill Net Ban," GCCA informed any sportsman who may have spent the previous year on another planet that "Florida conservationists are expected to succeed in their effort to ban gill nets and other entanglement nets from their state's waters; the vote there is Nov. 8, 1994. Alabama's Legislature is also expected to pass a 'No Nets' bill during its spring 1995 session. Danger will be on the horizon as outlawed Florida (and Alabama) gill netters begin making their way here. The number of gill netters in Louisiana could swell tenfold."

CONVINCING THE PEOPLE

LACA also got into the action, with a full-page ad: "ACT NOW! Preserve Our Coastal Resources Before It's TOO LATE!" ... "Yes, you CAN completely fish out a species. Many fish species are in danger of extinction."

Both groups solicited members and money. Joining LACA won the new member a gift subscription to *Louisiana Sportsman*.

Immediately after the November vote in Florida, the mainstream press got into the act with an Associated Press article that appeared in newspapers across the state. Captioned "Florida fishermen may flock to La. as nets banned," it read, in part:

> "Florida's new ban on most fish nets could send more out-of-state fishermen to Louisiana waters, environmentalists and others said Wednesday. Florida voters Tuesday outlawed fishing nets, called gill nets, from all Florida waters.... Florida fishermen may seek to move to Louisiana or other coastal states as a result, which could spell disaster for Louisiana's coastal resources, said Jeff Angers, executive director of the Baton Rouge-based Gulf Coast Conservation Association. 'If only half of Florida's gill netters make the trek to Louisiana waters, the potential damage to our marine resources is mind-boggling,' Angers said.... Angers said California banned gill nets or entanglement nets in 1990, Texas in 1988 and South Carolina in 1986. Mississippi and Alabama are expected to follow suit within the next nine months, he said. 'This is frightening to conservationists for fear that stocks would be depleted,' Angers said."

The AP article also mentioned the fishermen's proposed moratorium, and quoted agency biologist Harry Blanchet: "It's hard to say what [Florida] people will do. Some will probably find a different job.

Some will retire, get out of fishing entirely. Some will adapt and fish under the new regulation. And, some will move. Where, will be another question."

The scientist was apparently not hysterical enough for the AP, which gave Angers the last word: "In Venice [Louisiana] right now you can't get a hotel because they're booked up for the next two months with Florida gill-net fishermen. They saw this coming and came over early."

SINGLED OUT

The mullet run was in progress. Striped mullet were fine to eat but in Louisiana they were fished almost exclusively for their roe which was exported, mostly to Asia. The fish spawned from October into March, but the run peaked in November and December, and if fishermen wanted to lay aside some money before it petered out they had to get with it.

The biggest fish with the plumpest roe usually started down the Mississippi River first, which drew the top gun fishermen from as far west as Cameron Parish and as far east as Florida to the end-of-the-road community of Venice. The room and board ate you up but you could also hit it big in the river, if you had the nerve: "With a floating net, you can get some," said Golden Meadow fisherman Daniel Hutchinson, "if you don't get run over by a ship."

Galliano netter Gordon Kibodeaux had made the trip to Venice, but during a lull in the action, he trailered his skiff across the state to Calcasieu Lake, near the Texas line.

"The mullet run late in Calcasieu but Gordon went over early just

CONVINCING THE PEOPLE

to check it out. You never know," said Hutchinson. "We went over together the year before to check it out. We took a Lafitte skiff to sleep in, and towed over three 'well boats' [a specialized shallow-running boat that had a well for the outboard engine near the bow, which kept it away from the net that was deployed and hauled in over the stern.] It was blowin' a damned blizzard, and took us three days to get over there. We caught a few fish, nothing great, but it looked promising so Gordon went back."

Arriving just after the Florida vote, he wasn't welcomed by the local sport-fishing community.

Kibodeaux was raised in the Everglades, on Chokoloskee Island, where his grandparents ran a fish house. After the national park engulfed the area and put an end to most commercial fishing, he moved to Bayou Lafourche in the 1970s with his parents and grandparents, who opened a fish house in Leeville.

Now in his late 30s, Kibodeaux was one of the state's strongest finfishermen. With a wife and infant daughter to provide for, and with the "sports" breathing down his neck, he had a lot on his mind as he searched Calcasieu's waters for signs of fish.

"We were fishin' darned hard, tryin' to make the best of it," recalled Kibodeaux, "but I already wasn't in the greatest mood when the guys from over there started callin' me on the radio to tell me some woman was out there, goin' from boat to boat, askin' for me by name. I'd heard of her, but I didn't know the woman. So I was already hot when all of a sudden they came over and put a camera on me.

"They're about 20 feet away in a sport boat, holdin' a camera on me, and here's this woman screamin' at me about comin' over here from Florida and killin' all the mullet, and how there wouldn't be any left for the redfish and trout to eat.

LET THE GOOD TIMES ROLL

"I know I should've gone over there and talked real nice to 'em," recalled Kibodeaux, who instead lost his patience, accelerated his boat for a short distance, then "coasted it up and tapped theirs, just hard enough to let 'em know I was there. First I told her that 'her' redfish and trout would choke on the size mullet our nets picked up. And then I told her, 'Lady, I hope all your redfish and trout just roll belly up and die!'

"Then she and I got into it pretty hot. She said, 'That's a Florida boat isn't it?' I said, 'What, you can't read or write either? You can't see that's a Lou'siana registration?'

"We were goin' back and forth, and every once in a while this guy kept puttin' his camera up to his shoulder. I'd point to him and tell him to put that thing back down on the bottom of the boat—I know how they can twist anything with a camera. After he sneaked it back up for about the third time, I told him if he didn't take that damned camera off me I was gonna throw it in the lake, and he was goin' right in after it!

"But he had the sound on all the time and that's what they got me for—threatenin' that guy," said Kibodeaux, who was booked with assault, after Lake Charles television station KPLC-TV pressed charges against him.

"Later on, I had to call up the Lake Charles DA, to find out about my hearing, and I asked him, 'Are you familiar with my case?'

"'Am I familiar with your case?' he yelled. 'It's been on the TV every five minutes!'

"When I went in front of the judge, I told him that they came and put a camera on me, but he said, 'That's not what it looks like here.' He had the tape that showed me runnin' up to 'em. So I got a $500 fine and a year's probation."

CONVINCING THE PEOPLE

According to KPLC-TV environmental reporter Theresa Schmidt, "What had happened, Mary Poe had called us and said we should go out to the lake to do a story on all this influx of people who were coming over here because they were banning nets in Florida. She said she was concerned about the depletion of mullet, which other fish eat."

The Channel 7 news spot opened with Schmidt's narration, "Sports fishermen here in Big Lake [Calcasieu Lake] already think they are seeing more out-of-state netters since Florida has voted to outlaw gill and entanglement nets." She then interviewed a sport fisherman who said, "There are more and more netters...and we did not have this kind of netting until the vote, so we are assuming...they are from Florida." Then back to narrator Schmidt, "We approached this boat to see if it was from Florida, and got a hostile reception."

"Get that camera off me!"

"Though the boat had Louisiana numbers, guide Mary Poe says she's never seen it around here before."

"'I've never seen it before and I've been working on this water nine years. Never seen the boat or the guy either,'" said Poe who, with her husband, operated a hunting and fishing guide service. A board member of the GCCA, she was also president of the newly organized Louisiana Charter Boat Association which, with GCCA and LACA, was spearheading the net-ban drive. (The Louisiana Marine Trades Association, a group of outboard engine and boat dealers, also fought for the ban. No in-state environmental groups supported the legislation.)

LET THE GOOD TIMES ROLL

STEM THE INVASION? MAYBE LATER

While sportsmen filled the papers with their letters—"Gill nets should be banned in Louisiana," (11/10/94); "Time to ban gill nets is now," (11/14/94); "Gillnet problem," (11/17/94); "Louisiana, too, needs net ban," (11/21/94); "Fishermen must push net ban," (11/24/94) —the commercial industry struggled to implement its moratorium.

Louisiana already had a statute on the books that said, in effect, the state would not sell a license to a resident of a state that didn't offer Louisiana residents the same license. But since Florida's recently passed constitutional amendment wouldn't take effect until the following July, Florida fishermen had until then to get into Louisiana's fishery. So something had to be done, and fast.

With the Legislature not due to convene until the spring of 1995, the Louisiana Seafood Management Council presented its plan for a moratorium to the Wildlife and Fisheries Commission at its November meeting: For the next five years, *no one* would be able to purchase a saltwater net license unless they held a license in at least three of the previous five years. Once the moratorium was in effect, the industry would develop a permanent limited-entry plan.

But the Commission balked when its recreational majority said the moratorium could be considered an illegal restriction of interstate commerce. The panel directed its attorney to research the matter, and be ready with an opinion for the next meeting.

At its December meeting, the Commission's attorney told the panel that it had neither the constitutional nor the statutory authority to implement a moratorium on the sale of a specific license, because the laws that gave the Commission its authority never specifically mentioned the word "moratorium."

CONVINCING THE PEOPLE

Letters to the editor: Knowledge and experience in fishery management not required.

Representative Odinet then produced a letter from the attorney for the House Natural Resources Committee that claimed the Commission did have the authority because the moratorium treated all fishermen the same and therefore didn't discriminate against non-residents.

GCCA countered with a letter from the attorney for the Senate Natural Resources Committee that said, no, the commission did not have the authority.

All public comment, including that from the influential Louisiana Wildlife Federation, supported the moratorium, but some of the sporting commissioners said they couldn't vote for it because taking action contrary to their attorney's opinion would leave them personally open to legal action.

Finally, the panel skirted the issue by voting 4 to 3 not to hold a vote on the resolution because state law required a 72-hour public notice before a vote, and the item on the agenda had called only for an update by the panel's attorney.

Setting their sights on the January meeting, the moratorium's proponents hoped to settle the issue with an opinion from the state's attorney general. The December meeting, however, left the impression that the sportsmen, who'd been clamoring for a year about an invasion of "5,000 netters," really didn't want to solve the "problem."

After the December Commission meeting, GCCA took a battering in the press, with the *Times-Picayune*'s outdoor columnist provocatively referring to the group as "pro-net," for lobbying against the industry's moratorium. "The group feels an influx of gillnetters will help its cause for a total net ban," he suggested.

CONVINCING THE PEOPLE

Angers responded in a December 11, 1994, letter to the *Picayune*'s editor: "I must make one point perfectly clear to your readers: GCCA believes that it is vital that all gill nets and entanglement nets be banned from Louisiana waters. To accomplish that goal, GCCA will introduce legislation during the regular session of the Legislature in March. With that point now clear, allow me to explain the moratorium issue...."

Because the moratorium discriminated against out-of-staters, it would likely face a lengthy challenge in the courts, and political action would fizzle while the legal battle was going on, said Angers, because "prudent" legislators would take a "hands off" approach. Yet, "Our fisheries resources are currently being over-harvested by gillnetters to the point that this vital natural resource is on the brink of disaster. And a moratorium on the issuance of new licenses does nothing to reduce this current fishing pressure.

"Many of our southern sister states have banned gill nets. But with the recent gill net ban in Florida, we now have commercial fishermen from that state rushing to Louisiana to use their indiscriminate nets here....Given such a situation, GCCA feels that it has an obligation to its members and to the coastal resources for which we've fought for 11 years to reject a band-aid solution and demand a final cure for the ailment. That cure is a total ban of gill nets, not a moratorium on new licenses."

With no truce for the holidays, New Orleans seafood dealers Sal Piazza and Preston Battistella co-signed a Christmas-eve letter titled, "Moratorium on gill net licenses is fair solution."

"A moratorium...not only prevents new licenses for gill nets from other states but also limits licenses to residents, putting a cap on the number of gill nets allowed in Louisiana once and for all. It will also allow those who have made a living fishing here from generation to generation to continue to do so. At present, commercial fishermen are only allowed one million pounds of trout and no catch of reds while sport fishermen last year alone caught 5 million pounds of trout and 7 million pounds of redfish. The nets are therefore used almost exclusively for mullet, sheephead and [black] drum, which are less important to sport fishermen, and some of which cannot be caught by conventional sport fishing methods....This moratorium and a limited entry fishery with individual tags should prevent overfishing of our fisheries, and begins the repair of the relationship between sports and commercial fishermen, which is for the good of the fishery....If we can implement a gill net license moratorium, all folks who enjoy the benefits of our state's resource can continue to do so."

Also on Christmas eve, the New Orleans Saints finished off another less-than-victorious season with a 30-28 win over Denver.

In the 25 years since the team's first season in 1967, the Saints had never made it to a Super Bowl. Some years, the perennially losing team's performance was so dismal that humiliated fans attended games with paper bags over their heads, lest they be recognized. Finally, in the 1992 season, the team showed some promise, with a winning season of 12 victories and only four losses. The fans' hopes were dashed the following season, however, when their team finished an equitable though unsatisfying season, with eight wins and eight losses. The 1994 season proved even more frustrating, with only seven wins and nine losses.

On New Year's Day, 1995, in a letter to the *Times-Picayune,* an angling attorney from Shreveport, of all places, dismissed the mora-

CONVINCING THE PEOPLE

torium as inadequate because it only addressed new fishermen. "What about the gill netters who are currently licensed? The proposal will not affect them at all, and they are already doing enough damage by themselves without any outside help.

"...Why did they pick this time to offer the deal they did? They know their days of unfettered plunder are nearing an end, and they wanted five more years to squeeze out the last dollar that they could.

"...The issue is plain—no entanglement nets in saltwater in Louisiana! That is the battle to be fought and won. No appeasement less than this will better serve the resource."

The state's attorney general wasn't ready with his opinion on the moratorium by the Commission's January meeting, but in February he gave it a thumbs-up and the commissioners finally directed the department to halt the sale of saltwater gill-net licenses.

At the same meeting, agency biologists revealed that a redfish escapement rate of at least 70 percent—more than twice the minimum recommended by federal biologists—had been achieved for the fifth consecutive year. In response, Commissioner Tee John Mialjevich again proposed a commercial harvest, and offered a plan that was based on one of the department's allocation scenarios. "We can increase the recreational bag limit from five to seven, allow a commercial harvest of one million pounds and still have a conservative spawning class escapement rate of 50 percent," said Mialjevich, who also pushed the industry's individual quota plan and improved bookkeeping systems to track the commercial harvest.

The recreational majority nixed his proposal because, they said, there was insufficient biological data and "no plan" to monitor the commercial catch, to which Mialjevich responded, "We can't control

800 commercial fishermen yet we can control 300,000 recreational fishermen?"

The sportsmen on the panel then passed a Commission resolution stating that no change should be made in the state's redfish harvest until a review was conducted by the federal Gulf of Mexico Fishery Management Council's Red Drum Stock Assessment Panel. Translation: "When Hell freezes over."

Nevertheless, fishermen planned to push their redfish package through the Legislature, which retained oversight over the Commission. More urgently, they had to forge a limited-entry program—which effectively closed the fishery to new participants—and win its approval as soon as possible after the Legislature convened in late March.

But by that time, a cascade of events rendered the netters so unpopular that the politicians virtually ignored them, and instead of giving the fishermen redfish, they gave them the axe.

AS SIMPLE AS ONE, TWO, THREE

Winning campaigns is all about The Big Mo, and building a steamrolling momentum proved as easy for the Houston-based GCCA as ticking down a to-do list.

The group formally kicked off its campaign when the Baton Rouge Press Club invited its legislative champions to make a presentation at a meeting in late January 1995. "The Bankston/Triche Marine Resources Conservation Act of 1995" would be introduced in the upcoming session, announced the bill's namesakes.

Physically imposing, Representative Warren Triche appeared to

CONVINCING THE PEOPLE

relish his role as the sportsmen's flag bearer, and wasn't above referring to the family fishermen as "criminals." Senator Larry Bankston was more polished. An attorney from Port Hudson, just upriver from Baton Rouge, he was a second-term state senator, a shoo-in for the third and, feasibly, a future gubernatorial candidate. His district encompassed East Baton Rouge, which included the State Capitol, and enough chemical companies and refineries to rank the parish as the seventh most polluted of Louisiana's sixty four. Carrying the sportsmen's anti-commercial-fishing banner therefore offered little downside from Bankston's home turf, and if the ambitious senator was looking to gain statewide recognition, well, he would.

"The reason so many other states have banned these nets is because they are indiscriminate killers of every kind of fish," Bankston told the reporters. "Fish caught in these nets are either mutilated or die...and when deployed the nets don't know what fish is in season, what fish is out of season, which fish is marketable or which species is endangered."

Representative Triche told the press that the primary function of the bill was to prevent the future depletion of Louisiana's marine resources, and to conserve both commercial and recreational fishing. "Once a certain fishery collapses, it is already too late. It could take decades or a generation to bring back that particular species.

"The bill does not reallocate any species to any one user group," Triche reminded the media. "Commercial fishermen will still have their quotas and limits of all the species they have today."

In a fact sheet distributed by the politicians, journalists were told that the gill-net ban would have little impact on commercial fishermen, who would be able to "fill consumer demand for the affected finfish" by using rods and reels.

LET THE GOOD TIMES ROLL

The fact sheet also emphasized the net fishery's relative insignificance, in terms of its number of participants and economic contribution: The state's "estimated 500 full-time resident, commercial gill-net fishermen" comprised only a "tiny portion" of the state's total 16,000 commercial fishermen, and the dockside value of their edible finfish harvest totaled just $20 million. Yet there were an "estimated 300,000 saltwater recreational fishermen in Louisiana. Their activity generates $1.25 billion annually for the Louisiana economy."

Fishery biologists within the state's universities and management agency had been in unanimous agreement that the fisheries were in a healthy condition. GCCA's fact sheet now introduced a lone dissenter: LSU's "Dr. Richard Condrey....believes that if nothing is done to reduce the number of young speckled trout caught in Louisiana that trouble lies ahead—trouble for the resource and the fishery, both commercial and recreational."

Five days later, GCCA's executive director gleefully notified the group's members, "It's official. The battle is joined:

"....Nearly every media outlet in the state attended the news conference held by Sen. Larry Bankston and Rep. Warren Triche...The press response was gratifying. I hope you got to see some of it....But we still have a long road to travel....Your own participation will be critical to our success, and it's not too early to start talking up the entanglement net ban among your friends and even to start contacting legislators.

"We know the legislators will be hearing from those who are clinging to the status quo. Wouldn't it be great if they first had a barrage of calls and letters asking them to vote for the gill-net ban?

CONVINCING THE PEOPLE

At the January 23 press club meeting, Rep. Warren Triche showed a picture of a redfish caught in a net while Sen. Larry Bankston looked on. Louisiana's redfish had been netted for generations but after the gamefish law allocated the species exclusively to anglers, such images drove them wild. *(Bill Feig/The Advocate)*

"We have a brand-new tool for you, prepared just for this battle. It's GCCA's own legislative directory called *How to Contact Your State Legislators.*

"I am enclosing your own copy of the guide. Treat it as you would *a valued weapon,* because that's what it is. It gives you the name, address and phone number of every member of the Legislature. It points out which members hold key positions of special influence on this issue (and need to hear from you regardless of who your own legislator is).

"And it even outlines some of the key points you will want to make in explaining the issue to legislators, to friends, and in letters to the editor of your local newspaper.

"I've prepared this booklet to better equip you for the battle, but let me caution you in the strongest terms:

"Do not...do not...expect others to take on this battle for you.

"Just because the state office of GCCA is gearing up for this battle...just because I am personally eating, sleeping, breathing this effort...do not assume GCCA or some random guy down at the Capitol will take care of preserving our marine resources for your grandchildren."

KILLING FLIPPER

Three days later—just over a week after GCCA formally kicked off its campaign—the Louisiana Charter Boat Association's vice president and his buddy, who happened to be a Wildlife and Fisheries enforcement agent, hauled a piece of gill net into their home port of Slidell, a rapidly developing recreational stronghold on the northeastern shore of Lake Pontchartrain. The net contained a dead cormorant, a couple of redfish, and a dead dolphin.

"When they brought that porpoise in, they brought it right in there to us," recalled a Wildlife and Fisheries marine biologist, who was working at the department's Slidell field office. "Of course, that's where the telephones were too, where they could show us and make noise, et cetera. They didn't make any calls from my office, but I'm sure they made some thereafter."

"I don't remember who called me, but it may have been someone at the marina or the guy who supposedly found it in the water," recalled the local newspaper's writer, Shanna Labourdette, who in turn notified a photographer. "I called the photographer and told her what a hot issue gill nets were, because she didn't want to go out there and

CONVINCING THE PEOPLE

take the picture. I remember Jeff Angers then wanted photos from our photographer. She was a single mom, needing money, and sold them to him really, really cheap, like $15 each. I told her they would have paid *much* more! GCCA sent someone to pick up the photos immediately.

"Partly because my husband was a member at the time, I called GCCA for a quote—no one from GCCA called me first, so I had no reason to believe at the time it was staged."

The February 1 *Slidell Sentry-News* article, captioned, "Dead porpoise inflames issue of gill nets," opened, "They've been called the most indiscriminate killers known to man, and a dead porpoise, a dead duck, and several dead bull redfish found in a gill net Tuesday proved the statement correct.

> "…The six-foot porpoise, which perhaps roamed the waters freely just hours before, was the object of sadness and disgust to a crowd which formed near Boudreaux's Marina, just outside of Slidell.
>
> "'This news saddens me deeply,' Gulf Coast Conservation Association President Jeff Angers said during a telephone interview late Tuesday. 'The discovery at Boudreaux's is one of the many reasons why the Legislature must take action to end this killing of our marine resources.' Angers said."

A follow-up article the next day acknowledged that "The graphic pictures of the dolphin presented in the *Sentry-News* Wednesday fueled fire for efforts to outlaw gill nets in Louisiana.

"The mammal was buried near Boudreaux's Marina, after a sad and disgusted crowd mourned its death. 'Imagine how many more have perished,' Angers said."

The article concluded, "A bill authored by Sen. Larry Bankston (D-Port Hudson) and Rep. Warren Triche (D-Thibodaux) seeks to ban entanglement nets, in an effort to preserve commercial and recreational fishing."

The timing of the dolphin incident—and its similarity to another that had recently been used to sway voters in Florida—aroused the suspicions of those in the state who were familiar with the fishery and the tactics of GCCA.

For instance, the Wildlife and Fisheries biologist who'd witnessed the landing of the dolphin's entangled corpse, had been in charge of sampling the surrounding waters, with nets, for nearly 30 years: "I can't speak for the biologists working in other areas of the state, but I can say this: we did not capture any porpoises in our study area for the period of time I was there. So that's kind of the proof in the pudding right there, if you ask me."

Bottle-nosed dolphin were abundant within Louisiana's shallow inshore waters, yet the possibility of entangling one in a net ranged from highly improbable to not at all, depending on how fishermen deployed their gear.

"Strike netting," which generally required the most skill, involved spotting concentrations of fish and encircling them with the net, which was then almost immediately hauled back into the boat. Not only would the presence of a predatory dolphin disperse any schooled fish—thereby deterring the fisherman from setting his net in the first place—but if one were inadvertently encircled, the strike netter would be on hand to immediately release it.

"Set nets" were gill nets that were anchored out to intercept passing

CONVINCING THE PEOPLE

Déjà vu? When the Florida Conservation Association's leader, who was also publisher of the popular *Florida Sportsman* magazine—officially launched the campaign to ban nets in Florida, the March 1992 issue of his magazine featured this double-spread photo of a Florida Marine Patrol officer pulling in a dead dolphin that appeared to have been killed in a gill net. A Collier County Sheriff's deputy later revealed that the mammal had in actuality been killed by a blow to the head from the propeller of a powerboat. After its bloated carcass washed ashore, on tony Marco Island, net-ban proponents concealed the wound with the net and contacted the local newspaper whose photographer sold the photos to *Florida Sportsman*. When the instigators left the scene they took the net with them. *(Eric Strachan photo, from Florida Sportsman magazine.)*

fish, and periodically checked. State law required that fishermen remain in the vicinity of their nets so, should their gear entangle a marine mammal they, too, would be on hand to release it. If, however, a fisherman left his nets unattended—illegally—it was at least *possible* to entangle and drown a dolphin, though, in the words of a fisheries specialist with Sea Grant, it was about as likely as "finding a spaceship tangled up in your clothesline."

Yet GCCA immediately set about making that dead dolphin look like five thousand.

"'This most recent find is a striking and sad testament to what we believe is one of the major problems with gill nets—they are indiscriminate killers of everything that becomes entrapped in their web....Bottle-nosed dolphins are only one of the many types of marine life that are killed by gill nets," said Rep. Warren Triche, in a GCCA press release that ran word for word as an article in the February 5 *Alexandria Daily Town Talk,* and was captioned, "Dead bottle-nosed dolphin cited as reason for gill net ban."

> "'Lost gill nets (known as 'ghost' nets) are especially insidious because they can drift in the water for years, entrapping and killing marine life the entire time. Nearly all gill nets are constructed of material that is not biodegradable; while in the water they can entrap and kill marine life indefinitely, or until someone finds them and pulls them out,' said Triche.
>
> "Jeff Angers...said this latest incident involving the dolphin is not the first of its kind.
>
> "'It's sad, but true. Bottle-nosed dolphins and other marine life are slaughtered by gill nets. These killer nets are simply not compatible with our ecology. They destroy everything that comes within their grasp.'"

The article finished with a reminder to readers that "Because of recognition gained through the long-running television series, 'Flipper,' the bottle-nosed dolphin is one of the most popular and beloved sea creatures in existence."

The myth of "ghost nets" fishing forever and ever represented yet another example of the sportsmen's co-opting of the environmen-

CONVINCING THE PEOPLE

talists' melodramatic rhetoric for their own advantage. The term dated to the late 1980s when Greenpeace and other environmental groups campaigned to ban the use of the miles-long gill nets that Asian companies drifted in the high seas.

In Louisiana's inshore waters, nets were limited to a maximum length of 1,200 feet. Strike netters weren't about to lose their gear—the nets are expensive and time-consuming to make. On the other hand, unattended and inadequately anchored set nets, which were generally far shorter in length, could be swept from their moorings in bad weather.

As they moved through the water, however, they rapidly coiled end for end into a condensed spiral, and as their webbing picked up shells and other debris, their corklines and leadlines were pinned together until the net was soon reduced to a harmless glob that in the warm waters of the Gulf, where clean substrates were at a premium, was soon encrusted with barnacles, oysters and other reef-building sea life.

BAGGING THE LOUISIANA WILDLIFE FEDERATION

The photo of the dolphin in a gill net, with some redfish too, was powerful medicine, and surfaced frequently during GCCA's campaign. But when the emotional shot was flashed at a statewide convention of sportsmen, the tactic struck a nerve with a few of the state's more hardcore hunters.

The venerable Louisiana Wildlife Federation was a state affiliate of the prestigious National Wildlife Federation. A coalition of over 30 organizations, the Louisiana federation had a total of 13,000 members. In addition to its core of hook-and-bullet clubs, the federation's membership also included local environmental groups, wildlife biologists, even a few wild-resource harvesters like the Louisiana Trappers and Alligator Hunters Association and the Louisiana Natural Freshwater Catfish Association. Each year, before the Legislature convened, the group met to adopt its official position on upcoming outdoor issues. Prior to the federation's 56th annual convention, a pair of coastal commercial fishing groups joined, to add their input.

The federation had been incorporated in 1940, and over the years had become the state's single most influential voice on outdoor issues. The group's clout in the Legislature stemmed from having members in every parish, and from the respect it had earned by adhering to scientific wildlife management.

If the Louisiana Wildlife Federation opposed GCCA's net ban, the Texas-based group would face an uphill battle in the Legislature. Winning the federation's support, therefore, became a campaign in itself, which GCCA's director initiated even before the convention with a personal visit to each of its rod-and-gun clubs.

Held in the North Louisiana town of Natchitoches, the conference spanned the weekend of February 17-19. In addition to manning an informational booth throughout the convention, GCCA put on an illustrated presentation before the Fishing and Boating committee.

In the meantime, Barry Schaferkotter, of the Lake Pontchartrain Fishermen's Association, had been working the archers. Married to a Native American, Schaferkotter also happened to be a member of the Bayou State Bowhunters Association.

Like netting fish for food, hunting with bows and arrows had ancient roots; for its most passionate adherents the traditional activity surpassed mere sport to become a handed-down way of life. Faced with charges they were killing or crippling "Bambi" with their razor-edged broadheads, the archers were just as vulnerable to the same sort

CONVINCING THE PEOPLE

of simplistic and emotionally manipulative campaigning—by anti-hunting "non-profits"— that GCCA was employing.

So when Angers again waved the dolphin photo in front of the committee, and Schaferkotter whispered to the bowhunters, "See what I mean," they did.

But the archers' delegates had their orders: "You can vote for the ban, or you can abstain. You can't vote against it."

After a contentious debate, the fishermen and their many supporters tried to table the net-ban bill, but their resolution failed by a vote of 39 to 36. With the four votes allotted to their group, the bowhunters might have ended it right there.

Instead the committee approved GCCA's net-ban bill. However, it also gave the thumbs up to the commercial fishermen's legislative package, which included the implementation of a limited entry licensing system, implementation of the sales card program, and an appropriate severance fee on all seafood produced from Louisiana waters.

After that vote, an informational "Gill Net Ban Panel" was held for the edification of the general membership which, on the following day, was to ratify—or not—the committee's resolutions.

George Barisich, of the St. Bernard-based United Commercial Fishermen, and the marketing board's Karl Turner argued against the net ban, while Angers and Louisiana GCCA president James Jenkins, Jr. argued for their bill. Jenkins, a contractor who'd inherited the Baton Rouge-based Jenkins Construction Company, ticked off the Marine Fish Conservation Network's litany of fishery crises, from the cod in the Northwestern Atlantic to the 1930s "collapse" of Southern California's sardine fishery, and emphasized the importance of keeping Louisiana a "Sportsman's Paradise."

LET THE GOOD TIMES ROLL

After a sociable evening, the final votes were scheduled for Sunday morning.

At breakfast, when they opened Sunday's edition of the *Natchitoches-Times*, the federation's members were treated to a full-paged ad by GCCA, which had been paid for by *In-Fisherman*, a Minnesota-based freshwater sport-fishing magazine. Under GCCA's graphic photo of the Slidell dolphin, the caption read, "Today's Bodycount: 1 Dolphin, 1 Bird, 4 Reds, 1 Gill Net."

The federation's full membership ratified both the commercial fishermen's limited entry plan and GCCA's net-ban proposal. Though the Louisiana Wildlife Federation would not actively push for the ban, on paper it was in support, and that was all GCCA needed.

A lot of the federation's members were displeased about the outcome, but none more so than retired fishery scientist Dr. William Herke.

A WELL-INTENTIONED MISTAKE?

A field man, Bill Herke had worked for the federal U.S. Fish & Wildlife Service, and spent 32 years studying the life histories of fish and crustaceans that used Louisiana's marshes as their nursery.

Certified as a Fisheries Scientist by the American Fisheries Society, Herke held the highest rank (fellow) in the American Institute of Fishery Research Biologists. He'd also received the wildlife federation's 1987 Governor's Award, recognizing him as Conservationist of the Year.

A resident of Baton Rouge, Herke had been dueling with pro-ban sportsmen in the letters-to-the-editor section of *The Advocate*, where he countered their wild claims with rational information, and advocated for sustainable fishery management. A February 10, 1995, letter

CONVINCING THE PEOPLE

Retired U.S. Fish & Wildlife Service fishery biologist Dr. William Herke. An accomplished angler, who'd built his own sport-fishing boat, Herke tirelessly advocated for scientific fishery management and for the state's family fishermen.

had concluded with his recommendation, "...Louisiana should impose a severance fee on the commercial catch of the publicly owned fish resources. Limited entry, ITQs [Individual Transferable Quotas], and severance fees are coming into increased use in the more progressive jurisdictions and are the wave of the future in fisheries management."

After the Louisiana Wildlife Federation endorsed the ban, he wrote a lengthy editorial in the July 1995 edition of the group's official publication, "Louisiana Out-of-Doors," titled, "LWF Made A Mistake."

"I think this was a well-intentioned mistake. For many years the LWF has attempted to see that our wildlife laws are based on sound science; this time we did not. Rather, I think the vote for the resolution was the result of a concerted propaganda campaign by the net ban advocates.

"...We should all remember that PETA (People for the Ethical Treatment of Animals) and similar Animal Rights groups are trying to have hook-and-line fishing banned. And they are using tactics similar to those of the GCCA. When this battle gets hot, we sport fishers will need all the help we can get, especially that from commercial fishers. Consider also the importance of having the support of commercial fishers in the looming fight to maintain public access to traditional fishing waters. The non-fishing public can better understand the desirability of catching fish for food than they can for recreation.

"...We still have time to enact a moratorium on the sale of additional gill net licenses (resident and non-resident alike) until a limited entry program is in place for the saltwater gill net fishery. Limited entry is a complicated, but workable, system. Properly designed, it would eliminate much of the conflict between sport and commercial fishers.

"...A lot of misinformation about the gill net ban issue has been circulated around the state. If some of it has appeared in your local newspaper, I suggest you help set the record straight by showing this article to the editor and sports columnist."

Herke later testified in the Legislature against the net ban, and he and his wife Joan, a registered nurse, continued to write letters in support of the family fishermen—and science—for some years afterward. On the pages of the state's newspapers, the Herkes were the fishermen's most ardent third-party supporters, but not the only ones.

CONVINCING THE PEOPLE

FOOD WRITER TO THE RESCUE

In a February 23 article, *Advocate* food editor Tommy Simmons weighed in on the issue. Simmons liked to fish, enough to own a second home on the coast, but that didn't stop her from taking up for her targeted audience.

In previous fish fights, the award-winning food columnist had written articles that advocated sharing the resources with consumers. Her emotional testimony before a legislative hearing, in which she detailed the malicious harassment she'd received as a result of those articles, was instrumental in turning back a 1991 drive by the GCCA to make the trout a gamefish.

This time, she went out on a boat. The article that she wrote about her experience was the first in a series of educational pieces, which she introduced with: "Cookbook author and television cooking show maven Julia Child warned food journalists last year of the threat to the world's food supply from what she called 'food terrorists.' These food terrorists don't want to let others enjoy foods…our only defense is knowledge and experience with foods, she explained. Knowledge is gained by knowing where foods come from—besides the grocery store. This week, the FOOD section features the first story in an occasional series on 'From Sea to Table' and 'From Farm to Table.'"

For her story, Simmons accompanied Golden Meadow fisherman Mark Boudreaux, who'd written several letters to the editor of *The Advocate*. Boudreaux had been fishing commercially for ten years, and was also an offshore oil worker and supervisor for Chevron. Working "seven and seven," he put in a week on the rig, and spent his seven days off strike netting black drum from his 18 1/2-foot plywood well skiff.

LET THE GOOD TIMES ROLL

Simmons described her day with Boudreaux in a pair of articles, "Consumers caught in middle of gill-net dispute" and "Ignorance is enemy in the gill-net issue."

"There's a familiar saying about not understanding other people's lives until you've walked a mile in their shoes. I wouldn't want to walk half a block in the white boots worn by commercial fisherman Mark Boudreaux...if he always gets the treatment from recreational fishermen that I witnessed last weekend....There were two recreational fishermen picking up their bass boats as we got the boat ready. If looks could kill, all of us, and particularly Boudreaux, would be pushing up daisies this week.

"I've never witnessed such open resentment. Forget the fact that my husband had a camera around his neck, and I had the camera case and note pad in hand. Our group was launching a commercial-style boat with two nets in the well. We were the enemy of the fishery, in the eyes of the recreational fishermen at the launch that afternoon.

"Folks, if my experience was typical on-the-water treatment of commercial fishermen, then the enemy isn't commercial fishing or recreational fishing—it's ignorance."

While describing how the fishery was actually conducted, Simmons countered some of the net-ban proponents' campaign rhetoric.

"...the fish pulled in weren't mutilated. The game fish caught in the webbing were gingerly pushed through the webbing and released without damage to fins or gills. They swam off the minute they were returned to the water....The webbing also didn't catch as much fish as I had been led to believe. The fish caught were uniform in size, a size targeted by the size webbing used in the net."

She also let Boudreaux have his say: "'I've got problems with some of the figures being quoted by Sen. Bankston and Rep. Triche and the

CONVINCING THE PEOPLE

recreational lobbying group, GCCA. First, they keep claiming that Louisiana is being flooded with commercial fishermen from Florida.... The total number of saltwater gill-net licenses in Louisiana in 1994, both resident and nonresident was 1,144. Of those, 1,017 were resident saltwater gill-net licenses and 127 were nonresident saltwater gill-net licenses. These numbers are up," he said, "but no more so than comparable increases in recreational saltwater fishing licenses as well.

"'In 1993-94, there were 265,759 resident saltwater recreational fishing licenses, up about 20,000 from the year before, and 33,896 nonresident saltwater recreational fishing licenses, down about 6,000 from the previous year, for a net gain of about 14,000 more recreational saltwater fishing licenses.

"'Secondly, Sen. Bankston keeps talking about mutilated fish being caught in gill nets. Does he think the dock would buy mutilated fish to try and re-sell to restaurants, many of whom buy whole fish? I just don't think any of these people have ever gone out with a commercial fisherman and seen how the net actually works.'"

BOUDREAUX AND THIBODAUX

Boudreaux was from Lafourche Parish, where French was still the first language among many of the older fishermen. Undereducated and shy about speaking in public, the parish's family fishermen had been represented by a progressive trade group but it had recently collapsed. Boudreaux singlehandedly tried to fill the vacuum.

After thoroughly researching the subject, he began firing letters off to newspapers, radio stations, legislators, and anyone he thought might listen. Eventually, he homed in on GCCA's Representative Triche, who lived near the Lafourche Parish seat.

LET THE GOOD TIMES ROLL

As GCCA ramped up its campaign to destroy Louisiana's net fishery, Chevron distributed the group's manipulative literature to its refinery workers. As shown here, Chevron also targeted the general public through its retail outlets.

Sixty miles inland and surrounded with sugar cane fields, charming Thibodaux was home to Nichols State University and—in marked contrast to the modest wood-framed houses of the fishermen and oilfield workers down on the storm-battered coast—grand plantation homes that housed some of the representative's most powerful backers.

Boudreaux had hammered Triche in a February 6 article in the Thibodaux-based *Daily Comet*, "Fisherman cites 'facts,'" where he countered each of the politician's published claims, and ended with,

CONVINCING THE PEOPLE

When the oil industry workers dredged canals to their well sites, they piled the dredged spoil alongside. The weight of the embankments impeded the subsurface tidal flow to the remaining natural marshland, which died off and reverted to less productive open water. Consequently, this area, crisscrossed by dredged canals, later became an open lake. The Texas Oil Company (Texaco) was the first major oil company to intensively work in the state's coastal wetlands. In 2000, Chevron took over Texaco. *(Walter B. Sikora)*

"Triche may receive notoriety for sponsoring this bill, but not with any honor from the commercial fishermen in the parish." (The paper gave equal time to GCCA's Jeff Angers: "'For too many years the Louisiana Legislature has paid homage to the seafood packers...And residents of Louisiana and certainly Lafourche Parish should be proud of a leader like Warren Triche. Triche wants to do what is right because he knows what he is doing is right.'")

Boudreaux then challenged Triche to a public debate—which

Triche declined—and was organizing a protest in Thibodaux when he got a call.

"It sure would be a shame for you to lose your job at Chevron. Better lay off Triche."

With discretion the better part of valor, he did, though he later sent a packet to each of the legislature's natural resource committee members: After calling for a "new paradigm," he told them, "Of course public support of this bill will be overwhelming as compared to the commercial response, we are outnumbered over 400 to 1. It is understandable that some of you may yield to the pressure of potential lost votes. Is the majority always correct? Surely the majority of voters in your district would be in favor of eliminating taxes, but would that be the correct thing to do? All I can ask of you is to do what's right, not popular. The future of many people rests in your decisions."

CHAPTER FIVE
THE HIGH ROAD

To clarify just what was the right thing to do, the non-partisan Louisiana Sea Grant College Program and the Louisiana Cooperative Extension Service sponsored an educational symposium at LSU's Baton Rouge campus, just before the Legislature convened. Federally funded Sea Grant and the LSU Cooperative Extension Service shared the same mission—to make scientific information accessible to the public, and to find practical applications for that information.

According to the natural resource economist who opened the session, the speakers at "Coastal Fishing '95: Finfish Facts," had not met before as a group. "We don't have a common goal other than one thing: To make sure we clearly speak about facts."

The speakers and their topics included: John Roussel, chief of the finfish section at the Louisiana Department of Wildlife and Fisheries, "Basic Fish Dynamics"; LDWF biologist Harry Blanchet, "Status of Stocks at Issue"; Cooperative Extension Service agent Jerald Horst, "Perspectives on Catches by User Groups"; Cooperative Extension Service agent Sandy Corkern, "Controlling Fishermen Numbers: Experience of Other States"; Cooperative Extension Service economist Ken Roberts, "Economics of Fish Harvesting"; and attorney Mike Wascom, of the Louisiana Sea Grant Legal Program, "Summary of '95 Proposed Net Ban and Limited Entry Bills."

Each of the presentations subtly built the case that sustainable fishery management was a tad more complicated than "Ban the Nets!"

As for the sportsmen's obsession with commercial fishing, agency biologist Roussel explained that fish populations were influenced by "hundreds of natural and man-induced processes" that "interact with each other, and change in response to each other.

"Although many of us would like to believe otherwise, it is a fact that the numbers and kinds of fish within a given waterbody are always in a state of flux.... That is a fact whether or not you even harvested because many processes are going on that affect the kinds and numbers of fish. ...The same processes would also happen if man never inhabited the earth. A look at historical records shows that species come and they go, and their position in the ecosystem changes with time because of all these processes that are going on."

Roussel concluded, "Finally, I think a basic understanding of the fundamentals of fish population dynamics is necessary if we are going to have intelligent and successful fish management. Otherwise, we are simply managing by throwing at a dartboard. You have to have an understanding of some of these aspects before you choose a management alternative."

Agency biologist Harry Blanchet followed his boss with a report on the actual status of a few of the state's more important fish. "These species provide both sport and commercial fishermen with harvest opportunities, and provide managers with a continuous source of controversy," said Blanchet. "Many species of fish are harvested by each user group every year in Louisiana, but we'll concentrate today

THE HIGH ROAD

on four species: black drum, red drum, spotted seatrout, and striped mullet."

The black drum, which had a life cycle nearly identical to that of the redfish, had a milder meat and was increasingly being targeted by netters. Though sportsmen didn't get worked up about the whiskered fish, and used to throw them back in disdain, after they learned that drum were good eating they kept each one that they incidentally hooked while fishing for reds or trout.

"…Based on the assessments to date, at the existing harvest rates, black drum stocks are very healthy in Louisiana."

As for redfish, Blanchet noted that department biologists had been catching an exceptionally high number of small reds in their trammel nets and seines. "The reason for the relatively high recruitment is not completely known, but environmental factors seem to be important," he said. The harvest of larger fish was also up, even higher than the managers had predicted. The elevated harvest was due to an abundance of fish, rather than an abundance of fishermen, Blanchet explained. The distinction was important, he said, because an increasing harvest that was the result of more hooks in the water, rather than an increasing pool of fish, could reduce the number of fish that escaped the state's waters to replenish the offshore brood stock.

"The stock of red drum is presently being harvested at a rate well below the standard requested of the state by the Gulf of Mexico Fishery Management Council," said Blanchet. "Personally, I believe that the abundance of red drum seen now in the estuarine waters of the state may be unprecedented, at least within the last 20 years."

Unlike the long-lived red and black drum, both of which could reach about 40 years of age, the spotted seatrout lived only five or six. Yet females could become sexually mature in their first year, when just ten inches long, and spawned many times over the summer. "The stock of spotted seatrout in Louisiana waters has shown strong resiliency," said Blanchet. "...Relatively recent regulatory changes, including the establishment of a 12-inch minimum size limit for trout harvested recreationally, a 14-inch minimum size for commercially harvested trout, and the establishment of a commercial quota to constrain harvest from that sector, have all acted to protect the stock of trout from harvest until they reach a size where they have had a chance to spawn at least once," said Blanchet.

Louisiana's commercial fishermen had long brought in striped mullet—9,000 pounds in 1930, for example. But the fishery didn't begin to take off until the late 1970s, when markets were developed for the roe in Europe and Asia. Landings had escalated since then but there was still plenty of room for growth, said Blanchet, if fishermen continued to use 3 1/2- to 4-inch-meshed nets, which targeted older fish that had already spawned at least once.

"At present we estimate the mullet stock in much of Louisiana is still under-exploited. Harvest rates in some areas may be near the rate which would maximize yield in those areas but other areas have very lightly exploited stocks," he concluded.

Jerald Horst, the state's senior extension service fisheries agent, was a fishery biologist and an avid outdoorsman who specialized in hooking trout. He defended the data that the state agency used to manage fisheries.

"I will give you an illustration of why I believe these statistics are good. Look at the speckled trout management that Harry Blanchet just described. He mentioned that we're managing this species on a 15 percent SSBR, which some researchers would have you believe is a real delicate line and very low. [Spawning Stock Biomass per Recruit,

THE HIGH ROAD

a technical measurement of a fish population's health, was the ratio of the total weight of mature fish in a fished stock as compared to the total weight that would exist if the stock weren't fished at all.]

"In 1989, while we were managing the fish on the basis of 15 percent SSBR, we had a massive freeze, which just devastated our fish population. If our fishery statistics were so far off that they were no good for management purposes, why, in just a few years, did our fishery population of speckled trout recover to today's level? In 1990, the year after the freeze, it was difficult to find a fish out in the marsh. Yet two or three years later, we have abundant trout. That's a very good illustration that the fishery statistics that we have in the state are very good and they serve us very well in management," said Horst, who added, "There is a big gap between perception and reality when it comes down to what is said by catch statistics, and what people believe about them.

"There is a widespread *perception* that the blackened redfish craze, which really started about 1983, stressed and encouraged over-harvest of this resource, creating the statistical hole in the spawning stock of redfish that you frequently hear about. Interestingly...it was in the year-classes of the fish that were spawned in 1975, '76, and '77 that exhibited the hole; and if overfishing indeed caused that hole, the overfishing occurred in 1979 to '81. So, the reality is that this craze probably didn't cause the statistical hole.

"...I believe that people don't understand statistics and choose not to use those statistics wisely. I want to emphasize that this is true of both recreational and commercial fishermen; it doesn't really matter which is reporting or discussing."

Cooperative Extension Service agent Sandy Corkern offered a broadened perspective on the industry's pending limited entry proposal. Corkern had facilitated a limited entry task force in 1989 when Louisiana's fishermen first considered such a program. After describing his experience with that failed effort, he explained that different problems warranted different solutions.

"It is therefore necessary to honestly and objectively figure out what the problem is. Problems may be related to biology (for example, when a user group has over-harvested or when species reproduction is affected by such things as pollution or loss of habitat); economics (when there are already too many participants in a fishery to make entry an economically viable employment alternative); or sociopolitical (when there's a conflict between user groups for a resource, either due to inequities in allocations, or when there is a perceived, but not real, overuse by one or more of the user groups)."

From a simple moratorium on license sales, to a full-blown limited entry program, with the catch divvied up into quotas for each individual fisherman, Corkern described the full range of techniques used to limit access to a fishery, and cited examples in California, Alaska, and New Zealand, which "took years to devise with a lot of participation by everybody. However, both in California and Alaska, these systems have been a very positive experience from the standpoint of fish managers, commercial fishermen, recreational fishermen, and the public in general."

LSU natural resource economist Ken Roberts began his session, "In my experience, splitting atoms is a lot easier than dividing fish!"

Roberts's father, Nash, was a beloved meteorologist whose steady demeanor during hurricanes had calmed New Orleans-area viewers for decades. Ken Roberts had put himself through school by guiding anglers, and a couple of his sons operated charter boats. Still, like the rest of the participants, he followed the highest road, which in fishery

THE HIGH ROAD

economics led to sharing.

"There are no non-commercial uses of fish," commenced Roberts. "There is nothing more noble about killing a fish with a hook and line than with a net. Economics are involved in both issues.

"Angling is important, but it is an economic system based on inefficiency. The more money we can get anglers to spend on fish without increasing the bag limit, the better off the economy will be....The commercial side is based on an efficiency system. You can't go out and repeatedly catch fish unless you are making money."

Trying to compare economic impacts—the amount of money sportsmen spent on their sport—with the amount of money that commercial fishermen earned by selling their catch for food, was an apples-and-oranges argument. Economists are concerned with measuring the true value of fish, explained Roberts, and economic impact figures don't do that.

"For example, the value of the ice you bought to go fishing can't reflect the value of fish," he explained. "It must reflect the value of that bag of ice because ice has different uses. If you don't use it when fishing, you might take it home for a party. If someone asked you the value of the bag of ice, you would probably say, 'Well, I paid a dollar for it.'...Therefore, the ice has got to be worth a dollar. If you assign that dollar value to the fish as well, you are counting the value of the ice twice. It can't simultaneously be the value of fish and the value of the bag of ice."

Cost/benefit analyses measured the true value of fish, explained Roberts. By graphing increasing allocations of fish versus the costs incurred in their harvest, he demonstrated that benefits rise only to a certain point, and then begin to fall. If, for example, commercial fishermen are allocated more and more fish, their benefits rise until they are producing so many fish that the price falls, which forces them to invest in bigger boats to haul in more fish. By superimposing the same sort of graph on the recreational fishery, the intersection of the two lines scientifically identified the allocation ratio that maximized the overall value of the fishery.

"The general conclusion of natural resource economists is that, strictly on the basis of economics, most fisheries that are heavily utilized by different user groups get maximum net benefits through some kind of [shared] allocation system rather than allocating all of the resource to one user group," said Roberts. Still, he reminded the audience, it would ultimately be politicians, not economists, who'd be deciding who got what.

Attorney Mike Wascom finished off the day's program with an analysis of the net-ban related bills: "We know there are going to be passions on every possible side of several of these issues because they will result in legislative decisions, and we will have to live with those decisions. ...House Bill 919 by Representative Triche already has 30 co-sponsors while Bankston's bill on the Senate side has 12."

GCCA videotaped the proceedings of the educational conference with the tacit understanding that the group would make the video readily available. But without ever mentioning it by name, the speakers had serially gutted each of the Texas-based group's arguments for taking the fish—so much for that video.

Instead, on the following day—the Sunday before the session

[Facing page] On the day before the start of the 1995 legislative session, GCCA ran this ad in the state's major newspapers.

THE HIGH ROAD

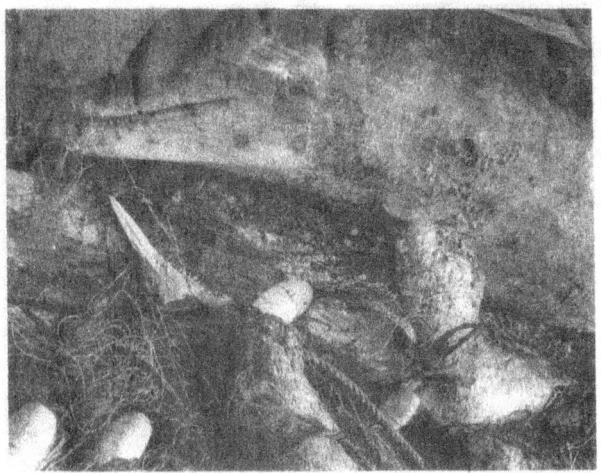

Today's Bodycount:
1 Dolphin, 1 Bird, 4 Reds,
1 Gill Net

Every day in Louisiana's coastal waters we have a body count of the victims of gill nets. Victims like the ones pictured above were discovered in an abandoned gill net near Slidell, Louisiana on January 31, 1995.

Commercial gill net fishermen say they use gill nets to catch fish only when they're in season and can be legally harvested and sold. The problem is the fisherman's gill net doesn't know what's in season or what fish or mammal is protected by law.

These giant curtains of death snare and kill about everything that gets entangled in their web, even dolphins, birds, and fish that are illegal to catch. Because gill nets are so destructive, Florida and Texas -- the two biggest Gulf Coast states -- have banned them from their coastal waters. Now some of Florida's 5,000 gill netters are headed for Louisiana.

But there's an organization in Louisiana that's fighting to save our marine resources from the ravages of destructive gill nets. It's the Gulf Coast Conservation Association (GCCA). Our members are Louisiana men, women and youngsters who enjoy fishing and who want to preserve our marine resources.

GCCA needs your help to save Louisiana's resources. Join GCCA today; it only costs $25. Or send a special tax-deductible donation of $100, $50, $25 . . . anything you can. We need you to join the fight for Louisiana's resources . . . before it's too late.

Gulf Coast Conservation Association
P.O. Box 373, Baton Rouge, LA 70821
Phone (504) 344-GCCA

Paid for by G.C.C.A., Jeff Angers Executive Director.

opened—GCCA ran oversized ads in several newspapers, including the Capitol city's *Advocate*.

In addition to the photo of the Slidell dolphin—"Today's Bodycount: 1 Dolphin, 1 Bird, 4 Reds, 1 Gill Net" —the ad debuted the group's new campaign logo. A knockoff of the one used in its Florida campaign, the graphic combined the slogan, "Save Louisiana Sealife!" with a stylized net that held a porpoise, a trout and a redfish.

The ad ended with a plea for tax-deductible donations. "We need you to join the fight for Louisiana's resources…before it's too late!"

CHAPTER SIX

IN THE LEGISLATURE

On Monday, March 27, more than 2,000 commercial fishermen, seafood workers, and their families marched on the State Capitol. After parading across the grounds behind a banner that read, "I Support A Louisiana Tradition: Commercial Fishing," protestors gathered on the Capitol's steps, where they displayed their handmade signs:

"Manage Fish by Science, Not by Politics"; "Help Us Keep Our Livelihood"; "Endangered Species, Commercial Fishermen"; "Our Net, Your Gain"; "Net Fishing Feeds My Family"; "Did Jesus Use a Hook and Line? Ban the Hooks"; "No Welfare. Let Us Work for Our Fair Share"; "Peter, James and John Were Fishermen. So Are We"; "God Gave Us Fish For Food, Not Just For Fun"; "God Said Cast Thy Net, Not Thy Rod-N-Reel"; "Don't let the GCCA Make Our Laws. Say No to Any Net Ban Bill!"

After meeting with the press, the protestors gathered in Memorial Hall, the cavernous lobby that separated the House and Senate chambers, and began to chant in a deep roar that echoed through the Capitol, "Eddie, Eddie, Eddie...."

Though he'd made a press appearance, where he held up one of the industry's theme shirts—"No Nets, No Seafood" —the governor didn't address the frightened fishermen. Instead, Senate President Sammy Nunez, who represented the southeastern coastal parishes of St. Bernard and Plaquemines, and was a close political ally of Edwards, expressed his support. The crowd whistled and cheered wildly as he vowed to fight the net-ban bill. Putting the fishermen out of business was criminal, said Nunez. "It's not fair. It's not right. It's not just!"

Statewide press accounts of the rally included quotes from fishing interests, biologists, and GCCA.

Preston Battistella, who distributed the fishermen's harvest to markets and restaurants from his facility in Faubourg Marigny, just outside the French Quarter, hadn't much cared for sportsmen since he was a kid: As he'd manned his Sicilian father's French Market seafood stand, it annoyed him when New Orleans anglers, looking for choice bait, grubbed through his shrimp in the pre-dawn darkness. And again when, after an unsuccessful trip, they returned to "Bayou Battistella" and rummaged through his neatly arranged fish to find the choicest specimens to take home. The anglers' subsequent attacks on his business hadn't improved his regard for them or their leaders:

"'If the resource were truly overfished, shouldn't these so-called sportsfishing conservationists also propose, in the interest of conservation, limits on the recreational take as well?' asked Battistella. 'They are nothing but fish hogs, cloaked in conservation garments. They want all fish caught on rods and reels. Since a rod and reel is not an effective manner of commercial harvest, the commercial fishing industry would be stripped of all real opportunity to participate in the fishery resource. Banning gill nets is not practical, not economical, and it certainly isn't necessary from a biological point.'"

Harry Blanchet backed up the latter point with, "'Regulations on

IN THE LEGISLATURE

On the first day of Louisiana's 1995 legislative session, fishermen and their families traveled to Baton Rouge from across the coast to demonstrate how many people a net ban would put out of business, in the present and in the future.

gill nets that affect minimum mesh size, maximum length, and restrict where they are used, generally prevent the over-harvesting of trout, drum and other fish.'"

GCCA's Jeff Angers stuck to message, contrasting the numbers of sport versus commercial fishermen, and inflating the value of the recreational fishery even beyond the oft-repeated and already bloated "economic impacts." The 500,000 recreational fishermen and their "investment" of $2.5 billion were at risk, he said, "if our marine resources are allowed to be ravaged by a small group of about 600 full-time, resident commercial gill-net fishermen."

The day after the Legislature convened, a woman from New Orleans wrote a letter to the editors of the *Times-Picayune* titled, "Gill netting must be ended."

"Gill nets must be stopped. The Louisiana Legislature is the place to stop them." If not, she continued, "There will hardly be anything left to protect, and Louisiana sports fishermen will become as rare as the fish they seek.

"...Earlier this year, a torn and discarded gill net was found with the rotting carcass of a dolphin (I'm talking Flipper here), a duck and several bull redfish. This net had been cast overboard after it had become useless.

"This net did not simply dissolve or go away; it had continued on its deadly mission, the entanglement of any and all living creatures in its path."

WAR CHESTS

The fishermen had their day at the opening of the session, but the

IN THE LEGISLATURE

net-ban bill wasn't scheduled for a hearing until May 3. In the interim, both sides built their war chests.

The industry's Louisiana Seafood Management Council hadn't yet established sustainable sources of funding. Urban dealers wrote the group checks, and as the council held meetings across the coast, fishermen and dockside buyers tossed cash into a cardboard box. In addition to *ad hoc* donations, they also collaborated in a "Two-Plus-Two" program where, each time he made a trip, the fisherman made a contribution of at least $2.00, and his buyer matched it. Except for oysters, however, late winter and early spring was the deadest time of the year for commercial fishing in the state's coastal waters.

Fundraising by GCCA was conducted at an entirely different level, as evidenced by a "confidential" April 6 appeal: "$2.5 billion. That's the amount of money invested by Louisianians today in durable goods related to saltwater recreational fishing (boats, camps and tackle). That figure doesn't even begin to address the trip-related expenses of saltwater anglers, i.e. fuel, groceries, etc.

"You yourself have contributed to this remarkable level of investment. Your passion for coastal angling led you to invest in it. We have also made sizeable investments in fishing. And we're fighting the battle to protect those investments from the onslaught of indiscriminate and increasing numbers of gill nets in our marine waters."

The appeal's introduction was signed by state chapter president Jimmy Jenkins, executive director Angers, and Chairman of the Board Jack Lawton, Jr., a Republican Party bigwig who owned an oil exploration and real estate development company near Lake Charles.

These leaders invited anglers to join them in the "Committee of 1,000." Membership was limited to donors of at least $1,000, yet, "Within the Committee, there will likely be members

['Commodores'] who decide they can support this effort at a level of a $5,000 donation. And a few exceptionally committed supporters ['Admirals'] who donate $10,000 or more."

As for "Who Can and Should Join the Committee," the list included "corporations desirous of a good image among conservationists; fishermen—and women—whose investments in the recreational fishery warrant the protection of the ban on gill nets, and who want to ensure the survival of marine resources for future generations; environmentalists interested in protecting Louisiana's natural resources; and anyone interested in preserving Louisiana as 'Sportsman's Paradise.'"

The goal was to raise $1 million, which would be dedicated to the conservation expenses of GCCA. In a breakdown of these "conservation expenses," however, the letter revealed that $100,000 had already been spent, "for the most pressing need: research and polling data," which included:

"Biological studies about the inability of our marine resources to sustain the current (and anticipated) levels of harvest....Economic studies about the importance of the recreational saltwater fishing industry....Socioeconomic studies verifying anew the exponentially greater value of a recreationally caught fish versus the same fish caught commercially. ...Statewide polls measuring support for such an 'environmentally friendly' initiative. The cost of each study and poll has been between $7,500 to $15,000.

"The next $50,000 is committed to the professional development of effective messages and communications tools that this highly contentious issue requires. The public's perception of this legislative initiative will, for the most part, guide the legislative arena.

"The balance of the funds raised through The Committee

IN THE LEGISLATURE

($850,000) will be dedicated to the public relations effort to educate the public on this issue. (Such educational messages 'trickle down' to legislators; they are members of the public too.) If appropriately funded, the prioritized public education effort will include: newspaper ads, radio and TV commercials, billboards and pamphlets encouraging a net ban and describing the irreparable harm caused by gill nets in general. Other items will alert Louisianians to the danger on the horizon brought on by Florida's success. Careful attention to detail will ensure that GCCA's tax-exempt status is not jeopardized."

POLLED BY GCCA

On April 18, Sen. Larry Bankston and Rep. Warren Triche released the results of GCCA's statewide poll in a news release titled, "Poll Shows Voters Overwhelmingly Favor Gill-Net Ban." Conducted by Baton Rouge-based Southern Media & Opinion Research, the poll sampled the opinions of 800 registered voters, 55 percent of whom possessed sport-fishing licenses.

Nearly 75 percent of respondents "disapproved of people who use gill nets to catch fish," more than 71 percent believed that continued use of gill nets would lead to depletion of fish resources, and 69 percent wanted their legislators to vote to ban gill nets.

"I have studied hundreds of polls over the past 20 years," said Bankston, "but rarely have I seen such a lopsided consensus of opinion on a single issue."

"It's clear that the people of Louisiana are not willing to trade away their precious marine resources for a few, supposed economic benefits," added Triche. "Voters don't want to compromise on this issue and they don't care who says a gill-net ban isn't needed—even if it's a scientist," he said, citing the 60 percent of respondents who favored banning nets even "if some state government fish experts and LSU scientists say we don't need to ban gill nets."

Bolstered by the polled respondents' 88 percent approval of recreational fishermen, Triche responded to Battistella's recent assessment of GCCA. "Commercial seafood interests have harshly attacked recreational fishermen, calling them 'greedy fish hogs,' among other things. Attacking recreational fishermen in Louisiana is like being against mom and apple pie. Voters want to keep Louisiana as the Sportsman's Paradise," he said. "These critics are hurting nobody but themselves through their vitriolic attacks on the 500,000 men, women and children who enjoy fishing in Louisiana's coastal waters, and who share their catch with family and friends."

The seafood industry countered GCCA's poll with a late-April press conference that included a few folks who needed a more reliable source of fish than donations by lucky anglers.

Lloyd Webber, of the powerful Louisiana Restaurant Association, predicted that a ban would raise the price of fish and cause a hard blow to the restaurant industry and tourist trade. "Restaurants have an annual economic impact of $3.4 billion in the state's economy....I think if this ban goes through, what you're going to see is supply that is limited and sporadic at best....Eighty-two percent of Louisiana consumers do not possess recreational fishing licenses. So where are they going to get their trout? They're going to go to restaurants."

Celebrity chef Emeril Lagasse added, "In Louisiana, particularly in New Orleans, there is a whole lot of tradition of dining, of food. When I look at the tradition of over a hundred years of dishes like trout *almandine* and trout *meuniere*, I can't imagine being in the 'city of dining' without these dishes."

IN THE LEGISLATURE

The press followed up the conference with a wave of articles that included quotes by other chefs and restaurateurs. "The restaurant industry is only asking for our fair share of the product," said Jerry Fein, who owned the Court of Two Sisters in the French Quarter.

Also interviewed, GCCA's Angers hit back at the restaurateurs: "I don't think the citizens of Louisiana are willing to have our marine resources indiscriminately depleted so a few fancy restaurants can make an extra 75 cents a plate on their $20 entrees. *National Geographic* called gill nets 'curtains of death.' That's exactly what they are. It's a type of fishing gear that has often been compared to fishing with dynamite, because it kills just as indiscriminately.

"Banning gillnets is not some sort of far-out reactionary measure. This is responsible preservation."

TAKE THAT!

Restaurateurs received swift retribution for speaking out. In a highly publicized April 20 sting dubbed, "Operation Bon Appetit," Wildlife and Fisheries enforcement agents busted several prominent chefs for illegally buying fish.

The year-long investigation began in March 1994, said Enforcement Division Colonel Winton Vidrine. "Two undercover agents posing as commercial fishermen began visiting restaurants and offering the businesses red drum and spotted seatrout. During each sale the agents indicated that the transaction was illegal," said Vidrine. "Additionally, anyone purchasing fish has the responsibility of requesting proof that the seller is a legally licensed commercial fisherman. All those arrested failed to request licensing information."

Of the 33 people busted, Chef Andrea Apuzzu was perhaps the most prominent and popular. A slightly built Italian immigrant, Apuzzu spoke with a rich accent and had worked his way up to owning the highly successful Andrea's Restaurant in the New Orleans suburb of Metairie. Chef Andrea was charged with three counts of buying red drum, three counts of buying from unlicensed fishermen, buying red snapper during closed season, and three counts of buying fish without a wholesale/retail license.

When he was arrested, he was handcuffed and paraded on television. The heavy-handedness offended many viewers.

"We are accustomed to watching television and seeing murder suspects handcuffed and being taken to Central Lockup. But watching the morning news April 20 took my breath away," wrote Marc Turk, in a letter to the *Times-Picayune*. "I have known and worked with both Mr. Apuzzo and Mr. Wileman [chef at Allegro's Restaurant in downtown New Orleans—buying speckled trout during closed season, buying from an unlicensed fisherman and buying fish without a wholesale/retail dealer's license] for many years. Everyone in this community knows the generous nature of these two men, and how they have given their time, money, hearts and compassion to every charitable organization that asks. To subject these two gentle men to such degrading acts as handcuffing and arrest is incomprehensible."

Turk had owned and operated The Bombay Club in the French Quarter for nine years, he wrote, but had never been told that he was required to ask for a seafood seller's license. "In speaking with several prominent restaurateurs (I dare not mention their names for fear they will be handcuffed and arrested) I found that they were not aware or, if so, only vaguely aware of such a law," he said, adding that "A letter to all restaurant owners outlining this law with its possible consequences would have been more understandable."

IN THE LEGISLATURE

THE MEDIA WEIGHS IN

In the April issue of *Louisiana Sportsman*, editor Todd Masson weighed in on the netting issue in a disturbing but clearly exaggerated account of a less-than-satisfying fishing trip with an anonymous fishing guide, that produced "only" 17 redfish:

"The pressing need of a gill net ban in our state has never been more real to the staff of *Louisiana Sportsman* than during a trip some of us took with a guide the last week of February.

"Due to several factors (some environmental, others perhaps not), we were only able to boat 17 keeper redfish, a handful of trout (only one of which was legal-sized) and five marsh bass. Slow days, of course, are part of fishing, but what struck me about the day was the condition of the fish and the deep concern of the guide about the future of angling in Louisiana.

"Of the 17 redfish we caught, eight had raw gill net scars around their necks, as did two of the bass. The passengers on board moaned in disgust each time such a fish was pulled from the brackish marsh.

"'That's nothing,' said our guide, who wished not to be identified. 'We've been catching them lately with green slime growing out of the sores. We throw them back because I think it's probably infection.'

"State statute requires that gill nets be constantly attended, but, as any boater knows, this law is often flouted by commercial fishermen. Even when nets are attended, they are left in the water for hours. During this time, ensnared fish fight fiercely for their freedom, often grinding the nylon nets through their protective scales.

"When a commercial fisherman finally picks up his net, and if he chooses to release a fish because it's not a legal species, that wounded, bleeding fish won't live very long. Multiply this by thousands of fish in hundreds of nets every day, and Louisiana's marsh is being steadily robbed of its future.

"To give us an example of the non-stop plunder of our state's fisheries, our guide took us to a remote bayou where he, only months before, had taken clients on a daily basis.

"'This whole shoreline was just loaded with redfish,' he said. 'You could see them everywhere.'

"Apparently the guide wasn't the only one with his eye on the reds. One day he took some clients to this honey hole and was disgusted to see a netter running his boat along the shoreline, driving the reds into an awaiting net.

"'We stopped and watched him unload his gill net,' the guide said. 'It was full of redfish.'

"The guide has returned to that shoreline nearly every day since the incident and has yet to catch another fish there. 'It takes an area a long time to come back after it's been fished out,' he said.

"Unlike the reds and bass, all of the trout we caught that day looked to be in good health; however, all but one measured only 10 to 11 inches.

"'That's all we catch around here now are those little bitty trout,' the guide said. 'I call them "filter trout" because they're the only ones that will fit through all the gill nets.'

"Few conservationists would be opposed to gill nets if they caught only the species they targeted and gave these fish some chance of escape. But, with deadly efficiency, gill nets kill nearly every fish that has the misfortune of swimming through their web. A netter targeting trout will kill redfish, black drum, sheepshead, flounder, croaker, marsh bass, mullet, catfish and a host of other species that play a crucial role in the food chain.

"The saddest part is that there's no valid reason for such waste. Commercial fishermen can catch the same numbers of targeted fish with rods and reels while releasing those that are undesirable or out of season. Like professional guides, they're

IN THE LEGISLATURE

on the water every day and thus will always know where the fish are. They may, of course, have to work an eight-hour day.

"There is no more crucial month for Bayou State sportsmen to help remove gill nets from our waters than April. The State Legislature, which will consider a gill net ban this session, convened on March 27, and many wavering lawmakers are waiting to see what direction their constituents push them.

"Call and write your legislators and tell them to vote for the La. Marine Resources Conservation Act of 1995. Urge your friends and family to do the same; this is a much more important vote than many of them likely realize."

Beyond the outdoor columns, newspaper editors began to take positions on both sides of the issue, generally being more for the ban the further their paper was situated from the water.

The influential *Times-Picayune* served the metropolitan New Orleans area, with a quarter of the state's population. Rather than take a stand on the question, the *Picayune* gave equal time to GCCA's Jeff Angers, and the Louisiana Seafood Management Council's lobbyist Manny Fernandez.

The two April 26 guest editorials were separated by a foot-high illustration of a fisherman; instead of proudly hauling in fish to eat, he was barefooted, enshrouded in a net with entangled starfish, his head hung low.

In "We must protect our resources," Angers drove home yet again his group's arguments, and ended with the emotional, "Louisiana is proudly known as the Sportsman's Paradise. For generations, mothers and fathers have enjoyed the tradition of fishing along our beautiful coast with their children—a tradition we cannot let down in miles of gill nets. How would we feel if one day our child or grandchild asked,

LET THE GOOD TIMES ROLL

"What was the Sportsman's Paradise?"

Fernandez, an attorney and a former state representative, had come from a Plaquemines Parish fishing family. With his roots in fishing, and his experience as a legislator, the seafood industry had hired him to fight the net ban. In his editorial, he reiterated the healthiness of the state's fisheries, and made an impassioned plea for fairness.

"In the final analysis, what is truly at issue is our legacy as citizens of this state and nation. All credible scientific and statistical evidence indicates that banning all entanglement nets in saltwater areas is not a reasonable or rational choice.

"Passage of such a law would create no greater economic, recreational or sociological benefit, even for those who propose it, while it would subject the commercial fisherman to financial disaster. Many commercial fishermen know no other manner of earning a living, sustaining their families or existing as productive members of society.

"The ultimate issue will be how we regard our fellow man. Do we believe that money and might are always right? Do we believe we may enrich ourselves by making others poor?

"Many commercial fishermen have endured much labor and hardship to care for, educate and prepare their children to become teachers, doctors, lawyers or public servants so they may be able to have a less difficult lifestyle. Most of those who share that legacy, like me, became recreational fishermen."

Fernandez ended his plea with, "The problem is not a gear problem; it is a people problem."

IN SESSION

On May 3, the same day legislators were to begin debate on GCCA's

IN THE LEGISLATURE

net-ban bill, the group ran a full-page ad in the Capitol city's *The Advocate*. Beneath a photo of a puppy-eyed child, the plaintive caption asked, "Daddy, what was Sportsman's Paradise?

"Wouldn't it be a pity if this little child's father were forced to tell his son that one time Louisiana was the Sportsman's Paradise? What if that father had to explain that gill nets had depleted our coastal resources, even though the vast majority of our people wanted our Legislature to ban these destructive devices, called "curtains of death" by *National Geographic* magazine?...The father would have to explain: 'In 1995, conservationists in Louisiana begged the Legislature to ban gill nets, but it just didn't happen because politics got in the way. Now we're not a paradise at all anymore. Sorry, son.'"

House Bill 919 called for an end to all netting in the state's coastal waters, and the creation of a commercial rod-and-reel license. For its first hearing, security was beefed up to the maximum. House Natural Resources Committee chairman Sam Theriot claimed that he hadn't seen so many cops at the Capitol since he'd tried to get the Legislature to ban most abortions.

Two committee rooms were set aside—one for the actual hearing and another with closed-circuit televisions that allowed additional people to watch the proceedings. Commercial fishermen, with their "No Nets, No Seafood" T-shirts, and sportsmen, with their beepers, overflowed from the two rooms, filled the hallways, and spilled out onto the Capitol's front steps where they smoked, glared at each other, and awaited—without incident—the outcome of the five-hour meeting. On cards provided by the committee, they stated their leanings on the bill: 350 were against, 250 in favor.

After the committee chairman laid down the ground rules for the hearing, four Wildlife and Fisheries biologists commenced their presentation.

Nearly 800,000 acres of the state's coastal waters were already closed to netting, either seasonally or throughout the year, they said. Grand Isle's beach was closed to netting in the summer, to accommodate visiting anglers; a mile-wide strip along Lake Pontchartrain's southern shoreline was permanently closed, to accommodate New Orleans anglers; Breton Sound, Vermilion Bay, federal refuges, the list went on.

The biologists then briefly described how the most popular fisheries were managed, and reiterated to the 19-member panel that the populations of trout, redfish, black drum, flounder, sheepshead and mullet were all "healthy."

When questioned about the threat of an invasion from Florida, LDWF Assistant Secretary William "Corky" Perret answered that, owing to the state's reciprocal license law, "We don't sell finfish net licenses to residents of Texas because Texas banned that particular license…and July 1 we will stop selling finfish net licenses to residents of Florida."

Perret allowed that the constitutionality of the law had never been challenged, and that he didn't know how many people might come from Florida in the intervening two months. An increase in effort, however, would largely be addressed by the existing management regime, he explained. The commercial trout fishery, for instance, was limited to a quota of one million pounds. Landings were tallied as the fish houses purchased the fish and when the quota was met, the fishery would be closed, irrespective of how many fishermen did the harvesting, explained Perret. "We do have management measures in place, i.e. seasons on some species, quotas on some species, net length

IN THE LEGISLATURE

and mesh sizes, and basically the mesh sizes are to allow our species to have spawned at least one time. Now, if indeed, we got an inordinate amount of pressure to come in from within or without, we may have to take some additional precautions, but at this time...we do have some restrictions that we feel are going to safeguard these species that we are talking about."

Committee members grilled the biologists for an hour and a half, when the chairman ended the questioning. "We have to legislate," said Theriot, who gave the floor to the proponents of the "Bankston/Triche Louisiana Marine Resources Conservation Act of 1995."

"I, like many, consider myself to be a conservationist," Triche told his fellow committee members. "My district, as is most of the state, is composed of a vast majority of recreational fishermen, who favor such a bill."

GCCA backed the bill, said Triche, as did the Louisiana Wildlife Federation and several other sport-fishing groups. Additionally, "At least right now there are 38 newspapers and magazines that are supporting this particular bill." (GCCA would later—in 1999—name Lafayette's Braxton I. Moody III "Conservationist of the Year," for "spreading the good news of good stewardship in his 33 newspapers statewide.")

Senator Bankston warned the legislators that, because the following session would be devoted to fiscal matters, they would not have another chance to deal with the issue until 1997. If the resource was depleted, their constituents would hold them accountable, and "you heard very clearly, very clearly, that the department cannot predict what is going to happen with Florida." If legislators chose to deal with "the onslaught" with a moratorium, it would not "withstand the impact of the Supreme Court," cautioned Bankston, who closed with the unintentionally ironic reminder that the resource belonged to the whole state, not just a few.

LSU associate professor Richard Condrey was the only fishery scientist to testify on GCCA's behalf. Though the proposed bill threatened the netters' ability to harvest more than 20 species of fish, Condrey limited his testimony to the spotted seatrout.

Agency biologists were managing the species "too close to the edge," he told the panel.

LDWF finfish chief John Roussel, in his presentation, had explained that the number of adult trout—a measure of the species' spawning potential—had remained stable over the years, so "the fishing pressure that exists right now can continue." The spawning potential "sort of fluctuates around 15 percent," he said, and by continually sampling with their nets, and updating their stock assessments, his biologists could anticipate any decline meaningful enough to require additional restrictions.

Condrey favored a more risk-averse approach. With a 20 percent spawning potential, "This fishery could be an ideal fishery. If we could move it into larger sizes, we could assure ourselves that the possibility of a decline in abundance would be diminished, and we have a good chance in increasing yield, because we are fishing and taking too many, too small fish before they have a chance to fully reproduce and to provide us with their bounty."

Under questioning, Condrey became increasingly emotional: "I feel like we're going down the road at 85 mph like we were doing with redfish, and I'm saying we need to slow down, and other people are saying...we haven't gotten into a wreck yet—and I say, 'Thank God'....

IN THE LEGISLATURE

Spotted seatrout harvested with 3 ½-inch-mesh gill net.

If the spotted seatrout stay at the current low level of spawning biomass, and never go into a state of decline, on the day I die, after I tell my wife and children how much I love them, I will say, 'Thank God.'"

In the thirteen years from 1980 through 1992, recreational anglers landed an average of more than 6 million trout each year while commercial fishermen harvested an annual average of slightly more than 800,000. In 1992, with a 3 ½-inch minimum mesh size in effect, the netters' trout averaged 1.76 pounds each, while the sportsmen's averaged 1.1.

Yet, no one proposed additional restrictions for the sportsmen. Instead, when a committee member asked the choked up professor if nets should be banned, he replied, "You are here to consider a net ban, you are here to consider the resource, I am here to express my concern for the resource."

GCCA president Jimmy Jenkins was more direct than Condrey: "Gill nets are indiscriminate, they mutilate," and they are "the instruments *National Geographic* calls 'curtains of death.'" The ruddy-faced contractor also told the committee that gill nets made enforcement of game laws almost impossible in a state with such a wide expanse of territory and a "permanent budget crisis" that prevented the hiring of a "sufficient number of enforcement agents.

"The enforcement division of the LDWF is among one of the hardest working agencies in state government. Having worked with them as a member and chairman of the Commission for a number of years, I can assure you that enforcement officers are woefully underfunded and understaffed and they cannot do it all....I don't know if it's lack of enforcement, ignorance of the law, greed, or what, but too many commercial netters are too willing to ignore our fisheries laws... Enforcement agents confiscate miles of illegal gill nets in Louisiana waters each month. The numbers for 1994 alone were 120,000 feet. In 1993, the numbers were 270,000 feet....If Louisiana conservationists are successful in banning gill nets from coastal waters, the job of our enforcement agents will be made much easier.

"During my years of service on the federal Gulf of Mexico Fishery Management Council, I heard and witnessed many horror stories relating to the declining fisheries. In fact, I witnessed this morning, the department here painting a very similar picture, 'Everything is okay.'

"I tried to cast my vote with principles of the conservationists and I tried to prevent nightmares from recurring. Nightmares like the cod

IN THE LEGISLATURE

groundfish fishery in Georges Banks, the halibut fishery in Grand Banks, the sardine fishery in California. All these fisheries were touted as healthy. All these were fisheries where commercial harvesters rejected conservation measures. All of these fisheries have died.

..."It's disheartening to read in newspapers and see Louisiana at the top of bad lists, like crime, and the bottom of good lists, like teacher pay. But native Louisianians are proud to be at the top of one list and that's the list of sportsmen's havens. In fact, we are the Sportsman's Paradise. The designation is worth standing up for and defending. And that's what Louisiana conservationists are doing in their efforts to protect our coastal resources from gill nets, before it is too late."

GCCA's "Committee of 1,000" had funded Texas A&M professor Robert Ditton to produce socioeconomic studies "verifying anew the exponentially greater value of a recreationally caught fish versus the same fish caught commercially," and he earned his money. In "A Social and Economic Assessment of a Net Ban in Nearshore Waters in Louisiana," he touted the net ban as a rural economic development tool.

Of all other Gulf Coast states, Louisiana had the fewest nonresident saltwater anglers, noted Ditton. If the state promoted its sport fishing, the tourist anglers' money would more than compensate for any losses caused by a net ban.

"For example: Quadrupling the number of licensed nonresident anglers, spending at the same rate as residents, for an average trip of around ten days, would generate total nonresident angler expenditures of around $121.6 million....This is nearly ten times the current dockside value of the entire commercial finfish fishery," said Ditton, who added that these expenditures would actually translate into a total economic impact of around $300 million. In comparison, the $14 million netters received for their product in 1994 was peanuts, and "could more than easily be made up with an emerging guide industry."

Because the finfish sector represented a relatively small portion of the seafood industry, a net ban would have little impact on the overall industry, said Ditton. And those business losses would be reduced by finding "substitutes for net-caught fish" such as "imports from elsewhere, fish caught by commercials using rod-and-reel gear, and aquaculture products."

In 1986, 97.9 million pounds of finfish, with a dockside value of $23.6 million were landed in the five Gulf states, said Ditton, while imports "were valued at $32.7 million. Thus over half, over one-half of the economic activity and produce markup associated with the wholesale distribution and retail finfish sector in the Gulf of Mexico is attributable to imports and hence is not affected by the current or any proposed net bans."

Recreational fishing's direct and indirect economic impacts overwhelmingly exceeded that of the commercial fishery, concluded sociologist Ditton. "About $621 million to about $39 million. Thus, if a finalization of scarce finfish resources were necessary, it would be in the state's best interest to allocate to the recreational fishery in order to protect these benefits."

"But there seems to me to be a culture and industry out there that has been depending upon commercial fishing for their livelihood," offered committee member Robert Adley, from Bossier City. If there was to be a gain of $300 million from tourism, there ought to be some money available to compensate the fishermen for their losses, and for retraining, he suggested.

Ditton agreed. When Texas put its fishermen out of business, "there was no concern expressed for social impacts" which, he felt, was

IN THE LEGISLATURE

"a mistake."

By now, Rep. Robert "Bobbie" Bergeron, from Houma, had heard enough. A French-speaking Cajun, Bergeron had grown up in the lower reaches of Terrebonne Parish, where his father had trapped furbearers and worked in a shrimp canning plant. Though he owned an oilfield service company, Bergeron hadn't forgotten his roots.

"By now, y'all have detected the accent. And let me tell you what is going to happen, besides the money and besides everything else. In the end, it will not be good for the whole. You're going to destroy a culture. You're going to destroy a way of life. You're gonna destroy people very much—very much like was done when the Native Americans were told 'We've got a good deal for you. We gonna bring you on some free land. By the way, we'll call it a reservation.' The overall society decided they needed to be moved from areas where they prospered as a people, and I see a parallel with the commercial fishermen and gill-net legislation that would retrain displaced fishermen so they could earn a living in a different way."

When the chairman called for amendments to the bill, he was bombarded with 25, and offered one of his own.

Apparently, of the 1,017 resident saltwater gill-net licenses sold in 1994, hundreds had been purchased by folks who set the nets out while they relaxed at their camps. So they could continue to do so, Theriot, of Abbeville, tried to create a 100-foot recreational gill-net license.

"It primarily affects a number of constituents that I represent," which included "two hundred families who have camps at Holly Beach," and who filed 200 cards in favor of the amendment. "I'm simply trying to take care of a group of people that I represent, just as you are, and many of you in this committee also represent the same kind of people," who likewise staked nets out around their camps on "Grand Isle, Cypremort Point, Rutherford Beach..."

The proposal from Theriot, who was backing the commercial net ban, drew a heated response from Rep. Jack Smith, a seafood-industry supporter from Patterson. "You're asking us to vote for a ban on nets that is going to put people out of work and which is going to cost the state a tremendous amount of money to reimburse those individuals, yet for a select few, we are now going to put on the bill allowing a legal net to catch the resource that they say they are trying to protect. Is that correct?"

The amendment failed miserably but its airing confirmed that the open-access fishery still included plenty of unskilled amateurs.

Kenneth Odinet offered an amendment to reinstate the fishermen's rights that HB 919 stripped away. Passed in 1991, the "Right to Fish" bill had set standards and policy for the management of the state's fisheries. Its language was based on the federal Magnuson Act's ten "National Standards for Fishery Conservation and Management," and essentially said that management decisions should be based upon science, and be "fair and equitable." According to the right-to-fish provisions, the Department of Wildlife and Fisheries was the public trustee of our fishery resources, and any action as sweeping as the proposed net ban required a study by the department. "The current law tells you, if you look at it, that you should not be entertaining this bill today," lobbyist Manny Fernandez explained to the committee.

The sportsmen had inserted the Standards and Policy bill into their net ban, and crossed out all references to the fishermen's "rights," and to "science and biology." Instead, future management decisions would be based upon "social and economic" factors.

IN THE LEGISLATURE

Odinet's amendment failed, and the sportsmen's brave new version remained in the bill to the end.

Rep. Bergeron offered an amendment to outlaw set nets, but allow strike netting to continue. Historically, the state's fishermen had employed trammel nets or haul seines. Because of the coarse twine used in their construction, these nets had to be set around schools of fish and drawn down until the fish were captured. After Florida fishermen brought their runaround monofilament gill nets to the state in the 1970s, locals—which included the South Vietnamese who, after the 1975 fall of Saigon, were being repatriated to coastal areas across the Gulf Coast—learned that the nets could also be staked out and left unattended. The highly visible practice infuriated sportsmen, who had quickly reacted by banning nets made of monofilament, and limiting the lengths of all nets to 1,200 feet.

Forcing netters to remain on their gear proved more difficult, however, because the definition of "unattended" was vague—the law at the time only required setnetters to be somewhere in the vicinity of their gear.

Since the very nature of actively worked strike nets bound them to the fisherman, allowing their use while completely outlawing the use of anchored nets "eliminates a difficult problem of enforcement," explained Fernandez. "Does it take away a tool from the commercial fishermen? You bet your sweet beep it does. But it also lets them live and continue their way of life and their culture and their traditions and their livelihood. It allows them to dress their kids and send them to school. It allows them to be a proud people.

"They do appreciate very much your concern for giving them compensation for displacing them, but they'd rather not be displaced. If you're gonna ban it—ban all their tools—then fine, the least you can do is give them compensation, but they don't want that. They don't want to be on your welfare rolls or be retrained. My goodness! People talk about retraining like a commercial fisherman could be retrained and perhaps become a PhD. who teaches at Texas A&M. Who we kidding? This is a decent way out, which allows you to ban the troublesome net. It takes away from the fishermen, but it doesn't kill 'em. It's a reasonable way to go."

According to his amendment, neither trammel nets nor seines would be affected, said Bergeron, because they, "by their very nature, are strike nets since they require the operator of the net to be on hand while the net is in operation."

If some members of the committee found the Cajun's discussion of nets dizzying, Triche brought it back to a level they could understand. "Mr. Bergeron, if I understand correctly, my bill is a 'ban-the-gill-net bill.'…So, the only thing that I can say is that you are either for the bill to ban all gill nets or you are for your amendment to reinstate gill nets in the form of strike nets. You either for nets or against nets, is the whole issue."

The strike net amendment passed 10 to 9. So did amendments to exempt the menhaden fishery and an already highly restricted pompano strike-net fishery around the Chandeleur Islands. Recreational anglers were disappointed when an amendment restricted the commercial rod-and-reel fishery to qualified commercial fishermen only. Another amendment established a $3.00 surcharge on saltwater sport-fishing licenses for two years. Monies raised were to compensate commercial fishermen for the loss of their gear, and fund their "re-education."

After five hours, the committee had little time to hear testimony from the bill's opponents, because of an impending meeting on the

IN THE LEGISLATURE

House floor: "I realize that the opposition was promised an hour, but with all due respect, most of the amendments offered, I think, came from that side of the track," said the chairman, "and if that wasn't sufficient, then you need to try and hurry."

The seafood industry had asked four specialists to offer technical information: Dr. William Herke, LSU economist Dr. Ken Roberts, and fishery biologists Charles Wilson, of LSU, and Dr. Dickson Hoese, from the University of Southwestern Louisiana.

In the fifteen minutes allotted to the bill's opponents, Herke briefly touched on the high points. Much had been made of the dolphin that had been found in a net, he said, but "This is a rare occurrence presenting no threat to the dolphin population. The National Marine Fisheries Service estimates that there are over 6,000 bottle-nosed dolphin in the Gulf, from the mouth of the Mississippi River to western Florida alone."

Yes, the Louisiana Wildlife Federation did go on record as being in favor of the net ban, allowed Herke, but barely. The vote "actually ended up being ten clubs for the ban and nine clubs against it."

As for Dr. Condrey's "worry about trout," Herke pointed out that "It's actually the sport fishermen that appear to be putting a bigger pressure on the trout and their reproduction than the commercials.

"And Dr. Ditton talked on economics, although he's a sociologist and I'm a biologist. Neither one of us are economics people. But I do know that the methods that he used were incorrect."

There was still time to enact a moratorium on license sales "until a limited entry program is in place for the saltwater gill-net fishery," concluded Herke. "....Properly designed, it would eliminate much of the conflict between sport and commercial interests. Along with commercial quotas on the various species, it could provide an equitable sharing of the fisheries resource on a sustainable basis."

Before they left, the committee members approved the amended bill without objection. The narrow passage of the strike-net amendment was a major victory for many, but not all of the fishermen. "Strike-net fishing won't work. If the agents enforce the law, it will not allow us to stake one end of the net out in the water," said Pete Gerica, of the Lake Pontchartrain Fishermen's Association. "We can't fish like that."

For the sportsmen, to whom "a net is a net," the amendment "gutted the bill," complained LACA co-founder Albert Bankston, a Baton Rouge printer and real estate developer, whose upscale recreational fishing development, Pointe Fourchon, was the first gated community in Lafourche Parish. "We still have the House floor and the Senate. We'll see where it stands after that."

The following morning, the House Natural Resources Committee also approved Kenneth Odinet's comprehensive netting reform bill. "This bill addresses the complaints that have been filed about gill nets, and takes into consideration the areas where most of the problems have occurred," Odinet told committee members. And "it allows commercial fishermen their God-given right to the fishery."

The compromise bill's major provisions included an immediate moratorium on new net license sales, and limited entry in the net fishery starting the following year. Only fishermen who'd held net licenses two of the last four years could apply for a license, and they had to prove that they'd earned either $5,000 or 25 percent of their income from saltwater finfishing during those years.

Netting would be limited to 180 days on the coast, and 120 days in inland saltwater lakes like Lake Pontchartrain. Increased commercial

IN THE LEGISLATURE

license fees were to be used for extra enforcement and implementation of an improved data collection system. Finally, the bill required the Department of Wildlife and Fisheries to study the impact of both net fishing and recreational fishing on coastal finfish populations, with the results to be presented to the Legislature by March 1, 1997.

Also on May 4, Senator Bankston presented his SB 412—a duplicate of Triche's HB 919—to the Senate's Natural Resources Committee. That panel agreed to allow strike netting to continue, but with additional restrictions.

The year-round mullet season was cut to just 90 days, during the spawning run. The trout season would coincide with the mullet season. During the combined season, fishermen could use nets with mesh sizes down to 3 ½ inches. The rest of the year, their minimum mesh size was 5 ½ inches, which allowed them to harvest larger species like drum, sheepshead, gar, and pompano.

On May 9, a thinly veiled fight over nets spilled over into the House Committee on Labor and Industrial Relations, when Warren Triche introduced his HB 2324, which threatened the bottom line of the restaurant industry, an opponent of his anti-netting legislation.

Virtually identical to a barter bill pushed by GCCA in 1981's Texas gamefish fight, Triche's bill called for the elimination of tips as part of the minimum wage, and forced restaurants to make up the difference. The measure would put some restaurants out of business and force layoffs of waiters and busboys, said Jim Funk, executive vice president of the Louisiana Restaurant Association. Restaurants were allowed to pay half the minimum wage to workers in anticipation of the tips they'd receive, said Funk, but if the employee didn't make the minimum wage with tips, federal law already required the restaurant to make up the difference.

"Why should we be getting in the middle of the restaurateurs' business?" asked Rep. Odinet, who was a member of the committee. "I have fishermen in my district who don't make minimum wage."

The committee narrowly approved Triche's bill, but its final passage wouldn't be necessary.

On May 16, GCCA again ran the schmaltzy "Daddy, what was Sportsman's Paradise?" ad in Baton Rouge's *The Advocate*. The group shared the ad's cost with the Federation of Fly Fishers, an international organization of fly fishermen headquartered in Bozeman, Montana.

When Triche had first aired his bill in the House Natural Resources Committee, the net ban's opponents had overwhelmed him with their compromise amendments. Now it was payback time.

When the full House met to take action on his HB 919, Triche and his allies showed up with eleven pages of punishing amendments. Before reading them, they put on a show-and-tell.

"Gill nets are the most indiscriminate fishing gear known to man," said Triche, as he and his colleagues awkwardly stretched a net down the middle of the House chamber. As if to make his point, he and the other greenhorns were repeatedly entangled as the webbing snared the buttons on their clothing.

With the curtain of death snaking down the aisle, Triche flashed photographs of net-caught redfish, trout, the Slidell dolphin, and reminded his peers that there were "500,000 sports fishermen and fewer than 1,200 commercial fishermen. Represent your constituency when you vote," said Triche, who, for good measure, presented each

IN THE LEGISLATURE

of the representatives with a list of saltwater gillnetters residing in their own districts.

His new amendments put GCCA back on track. The harvest of all fish except mullet would be limited to rods and reels only. Since mullet were vegetarians and didn't readily bite on hooks, they could be taken with nets, but the season would be even shorter than had been agreed upon earlier.

Until 1995, the mullet fishery had been a 365-day-a-year endeavor. During the roe season, which ran from the third Monday in October through mid-January, fishermen could bring in unlimited quantities. After the fish recovered from spawning, they became prime food fish, and were trucked to neighboring states, and as far off as California, where they were processed and exported to Asia. From the close of the roe season until September, fishermen were limited to 1,500 pounds per day. From September until the opening of the roe season, the fishermen's daily quota was reduced to 200 pounds, to protect the fish as they plumped up with roe. Additional regulations, including a minimum mesh size of 3 ½ inches, ensured the fishery's sustainability.

Sportsmen, however, claimed that the mullet had to be further protected because it was "part of the food chain," and the full House responded by cutting the mullet season from the 90 days that the House Natural Resources Committee had agreed upon, to just 42 days. The season would begin on November 1, no fishing would be allowed on weekends, when sportsmen might be on the water, and the minimum mesh size was raised to 4 inches, an impractical size in most regions of the state.

To obtain their mullet net license, fishermen were required to have had a saltwater gill-net license in any two of the years from 1992 to 1994, and had to verify with their tax returns that at least 50 percent of their income came from the sale of seafood products. The same qualifications applied to any fishermen who wanted to purchase the $250 rod-and-reel license. Fishermen had to choose between the license—they could either net mullet for 42 days a year, or use a rod and reel to hook the other species—they couldn't do both.

Since there were no provisions for the transfer of these licenses to new recruits, the number of fishermen would inevitably decline. When asked whether this could ultimately end all future commercial fishing for the state's estuarine finfish, Triche snapped that the fishermen "had no future."

The amended bill retained the $3 surcharge on saltwater recreational fishing licenses, but for three years rather than two. However, commercial fishermen who chose to obtain either the rod-and-reel or net license would be ineligible for any compensation from the Commercial Fisherman's Economic Assistance Fund. And only two-thirds of the money raised by the surcharge would be available for compensation; the other third would be set aside to hire additional enforcement personnel in the Department of Wildlife and Fisheries.

The amended bill also did away with the requirement that the department present its annual stock assessment and allocation scenario on redfish, effectively sealing the vault on that species. Instead, it required the department to prepare an annual report on mullet. If the spawning stock biomass fell below 30 percent, netting would immediately be stopped "for one year."

Among various additional restrictions, the bill created a $250 traversing permit that allowed fishermen to cross state waters if they chose to net in federal waters, which began three miles offshore.

Rep. Bergeron told the House that the amendments were the result of greedy recreational fishermen who wanted to end commercial

IN THE LEGISLATURE

fishing, to which Triche responded, anglers just wanted "to have a chance to go out and have a decent time, and at least catch fish once in a while."

"This is going to be devastating to this culture and the people who live in it," warned Bergeron. "This is a social justice issue in every sense of the word. Those kids in those areas get cold, they get hungry and, yes, they have ambitions to get educated," he implored. "This bill will be an instrument of death, but not to fish—to human beings. You're about to do violence."

"You know not what you do," added Fernandez. "We sit here in our Brooks Brothers suits and our linen shirts and our $85 painted ties, and we deal a death blow to a culture."

With one Representative absent, the House voted overwhelmingly—90 to 14—to approve GCCA's re-invigorated bill. The lawmakers then killed HB 2436, Rep. Odinet's moratorium/limited entry package.

A few days after that rout, Fernandez, the Louisiana Seafood Management Council's lobbyist, briefed disheartened fishermen and seafood dealers.

Before the final 90 to 14 vote, up to 40 of the representatives had backed the fishermen on some of the amendment votes, he said, but when it came time for the final vote, they saw that the bill was going to pass, so they voted for it "so as not to offend the recreational fishermen."

Since it looked as though HB 919 was going to pass, he would focus on amending it, "to decrease its harsh impact," Fernandez told the fishermen. Amending the bill didn't jeopardize the industry's right to contest the government's actions in court, so they had nothing to lose by continuing the fight, and might gain something.

"It is not a time to abandon our effort to fight this legislation but rather it is a time to join together even more closely," said Fernandez, who encouraged fishermen to contact their senators as the bill moved to that side of the Legislature.

"The message to those persons is the devastating economic impact on the commercial fishing industry, the dealers, the processors, and the dockside persons who work in the industry. We should also emphasize the impact to the consumers on higher costs and to the restaurant and tourism industry. This point should be emphasized by the fact that there is no biological information to justify the actions to displace this entire industry.

"Our noble cause is one worth fighting for, for the alternative is to surrender," Fernandez concluded. In spite of his pep talk, fishermen left the meeting like dead men walking.

PUBLIC DEFECTION

The May 16 spectacle in the House chamber had turned off one legislator. Though he'd been quick to sign on as a co-sponsor of HB 919, David Vitter was one of the 14 representatives who voted against the bill. A few days later, he very publicly defected from the Ban-the-Nets movement when he asked that his name be removed from GCCA's bill because there was no evidence that it was necessary.

A graduate of Harvard, and a Rhodes Scholar at Oxford, Vitter had graduated with honors from Tulane Law School, and was an adjunct law professor there and at Loyola University. A Republican from Metairie, Vitter was at odds with many legislators because of his stance against gambling, which they'd recently legalized. As an ardent

IN THE LEGISLATURE

advocate of good government, he was known to lecture fellow lawmakers on ethics, which didn't enhance his popularity. Indeed, Vitter was riding the mid-1990s tide of voter discontent with a bill in the 1995 session that sought to limit the terms of Louisiana's legislators.

Vitter announced his break with GCCA in a May 22 *Times-Picayune* article. "Quite frankly, I don't enjoy admitting this, but I didn't do my homework. I assumed there was significant scientific evidence that the resource was being depleted, and that gill nets were a part of that. I asked again and again for any scientific data supporting the ban," said Vitter. "Essentially all I ever got was that it was favored by 70 percent of the voters and how much more important the recreational fishermen are to the economy than the commercial fishermen."

Vitter's district included Bucktown, a unique urban fishing center with a handful of finfishermen who netted fish in Lake Pontchartrain and landed them into the city of New Orleans. His district also included thousands of recreational fishermen.

"Certainly, politically, it would have been easier to vote for the bill," he continued. "There's all sorts of anecdotal evidence that all the recreational fishermen have given me—and I don't discount that—they feel like their ability to catch their limit has gone down. But in the final analysis, I cannot justify having government take such a dramatic action, which would involve essentially putting a thousand people out of business, without very clear and convincing scientific evidence."

As they tried to get back on their feet, the fishermen highlighted Vitter's high-profile defection in their news releases: "I didn't do my homework...."

Though the House had killed industry supporter Odinet's limited entry bill, HB 2436, he had another one in reserve. HB 998 called for a three-year moratorium on the sale of new gill-net licenses, and limited the issuance of new licenses to those who'd held such a license in two of the past four years.

Odinet had already passed his bill through committee and, before the full House he explained that it would prevent a rush of out-of-state fishermen from buying licenses. Net-ban supporter Peppi Bruneau, a Republican from New Orleans, responded that the bill was designed to undermine HB 919. "This is a very clever piece of legislative draftsmanship," said Bruneau, because it would essentially repeal the ban if it were passed after Triche's bill cleared the Legislature because the last legislation passed would take precedence.

"This bill started out as a little minnow swimming though the legislative gill net," said Bruneau. "By the time it is finished it will be a barracuda."

"You may know a lot about video poker but you don't know anything about moratoriums," Odinet told Bruneau, who'd led the drive to legalize video poker in the 1991 session.

"If you want to keep people from Florida and other states from coming in and getting licensed, vote for this bill," Odinet told lawmakers.

Instead, the House voted 65-32 to kill his bill, eliminating any risk that it would derail GCCA's net ban.

The Senate Natural Resources Committee heard Triche's HB 919 on the morning of May 25. The seafood industry had few supporters in the Senate. Grossly outnumbered, Senate President Sammy Nunez, a Democrat from Chalmette, and Senator Marty Chabert, a Democrat

IN THE LEGISLATURE

from Chauvin, nevertheless won the fishermen a few concessions.

As it had passed the House, HB 919 cut the mullet season from 365 to just 42 days, and limited the commercial harvest of trout and other species to rods and reels. The Senate committee lengthened the roe mullet season to 90 days, and reopened the trout fishery to strike nets during the same season, with nets of at least 3 ½-inch mesh. The Senators also allowed the use of strike nets with a 5 ½-inch minimum mesh throughout the year.

The committee also slightly loosened the qualifications needed to obtain strike net and commercial rod-and-reel licenses.

The concessions wouldn't result in the net-free resort that sportsmen envisioned. "All the senators would do, if this bill passes as amended, is to fiddle with the gill-net laws," said GCCA's Angers. "We believe there is a demonstrated vote in the House—by 90 to 14—to do more than fiddle with the law," he told a reporter, adding that when the bill moved into the full Senate, his group would try to strip those amendments that allowed any netting to continue.

On Monday, June 5, the Senate gave final approval to HB 919 after two teams of senators announced that they'd reached a deal, which they hurriedly scribbled into a collection of amendments.

During the last-minute negotiations, coastal Senators Nunez and Chabert were assisted by an unlikely source—GCCA's Larry Bankston.

Bankston and GCCA had fallen out when the senator realized that Governor Edwards would veto a total ban. GCCA members had "an absolutely fanciful belief that they can override the governor's veto of this," said Bankston, who acknowledged that the group had a legitimate point in arguing that eliminating the use of all nets made it easier to enforce restrictions on unattended nets. Still, the bill that came out of the Senate committee, which allowed only strike netting, was a major improvement over the current law, said Bankston. Since he felt that it would also pass muster with Edwards, he helped the fishermen's supporters get it passed.

Without Bankston, GCCA tapped Senator Donald Cravins, from Lafayette, to help long-time GCCA man John Hainkel negotiate for the sport side.

Senator Hainkel, a Republican, was an attorney and a popular veteran in the Legislature, where he represented the interests of the Uptown and Garden District elite of New Orleans. Born in 1938, he told the Senate that shrimp and crabs had been plentiful on the New Orleans lakefront when he was a boy. "Now you're lucky if you catch a beer can."

"Technology" was to blame, said Hainkel. "Gill nets had outrun the fishing resource's ability to reproduce."

Scrambling to put together an acceptable "compromise," the two teams of senators worked right up to the vote. Their amendments led to a two-year phase out of all netting except for mullet and pompano, and just enough of that to avoid a gubernatorial veto.

Trout could be netted for two more years. By 1994, the commercial trout season had been trimmed from 12 months to 7 ½ months, from September 15 to May 1, or until the 1-million-pound quota was reached. The senators further clipped the fishermen's last two seasons, to less than 3 ½ months. The first season was to begin on November 20, 1995, and end no later than March 1, 1996. The last season was to begin November 18, 1996, and end by March 1, 1997. After that date, the commercial trout harvest would be limited to the same rods and reels used by recreational fishermen.

Qualified fishermen could continue to net roe mullet indefinitely,

IN THE LEGISLATURE

Lake Pontchartrain's southern shoreline, at New Orleans, was edged softly with productive marshland until the 1930s, when the Works Progress Administration and the Orleans Levee Board built the concrete seawall shown in this 1941 photograph. The wetlands behind were then filled in with spoil dredged from the lake. The "reclaimed" marsh was later developed as Lakeview, a residential area that was subsequently inundated by the lake's waters during Hurricane Katrina in 2005. Most of the city's untreated stormwater runoff also flowed into the lake in this vicinity, via the 17th Street Canal. *(Louisiana Division/City Archives, New Orleans Public Library)*

during a short season to run from the fourth Monday in October to the fourth Monday in January.

From a statewide beach fishery, pompano netting was confined to a limited area east of the Mississippi River, where a maximum of eight specially permitted fishermen could work. The long-standing special six-month season there was reduced to three months.

All netting was banned on nights and weekends.

With these amendments, the full Senate passed HB 919 by a 39-0 vote.

The following day, on June 6, the House voted 92-10 to accept the changes made in the Senate, and sent the much-amended bill to the governor. Edwards said he would sign it, with reluctance.

"If I were convinced that the fish population was decimated or damaged by any kind of net, I would support a total ban, but there is no evidence of that at all," said Edwards. "Of course it is going to hurt commercial fishermen and also impact the restaurant industry, but the sports fisherman is the order of the day."

The next morning, Senate staff attorney George Carmouche dropped a bombshell when he announced that the bill that had been sent to the governor wasn't the same as had been agreed upon.

"To call it a mess is an understatement," said Carmouche, a specialist in wetlands law. After spending what he described as a sleepless night reading the final version of HB 919, he concluded that, "It doesn't say what it was supposed to say."

Carmouche had been present during the Senate negotiations, and for corroboration, turned to Senator Bankston. The teams had agreed upon a minimum mesh size of 3 ½ inches for the mullet and trout season, not the 4 inches that was written in the bill, confirmed Bankston, who added that the bill's text mentioned only trout, mullet and pompano, while the intent had been to allow fishermen to continue to net drum, sheepshead, gar, shark and other species with a five-inch net for two more years.

Although Angers insisted to a reporter that the bill "does say what

IN THE LEGISLATURE

we intended to say," Carmouche's findings stunned net-ban proponents, who feared that the discrepancies could sidetrack their hard-won bill. Governor Edwards was already taking another look at it.

With the session due to adjourn on June 19, time was running out. The key players on each side of the issue met on the night of Thursday, June 8, to iron out the differences but failed to do so. After another meeting on June 9 also led nowhere, further negotiations were scheduled for Monday morning, June 12.

That will be the final meeting said Representative Triche. "It's time to get something done."

BAD BOY

When the legislators met on Monday morning, they weren't too happy with GCCA's executive director. On the Saturday before, Angers had been speaking at a GCCA meeting in Lake Charles. Situated in the west, near the nation's most polluted state, Lake Charles was noted for the demoralizing stench of its own petrochemical industry, which earned Louisiana its ranking as second-most polluted. The area's largest employer was a chemical plant owned by a Pittsburgh-based corporation that was being sued by netters after its emissions of PCBs rendered their harvest of trout and other species unmarketable. Predictably, the Lake Charles chapter was GCCA's most militant.

Speaking to that audience, Angers apparently got caught up in the moment: "We felt the compromise dealt mostly with bottom-dwelling, scum-sucking fish, but instead apparently it may indeed deal with bottom-dwelling, scum-sucking legislators that are not men of their word."

The comment had been publicized in the media before the Monday meeting, and brought an angry response from legislators in both the House and Senate.

"He's talking about us," said Senator Chabert. "This makes me sick. We never sat up here and talked about little rich, red-headed kids representing sports fishermen."

In his apologetic letters to the net-ban opponents, Angers said the comment "was made totally in jest...The remark does not reflect my feeling or those of the organization I represent toward the Legislature."

Senators Hainkel and Cravins also apologized for Angers. The comments did nothing to help resolve the difference between the commercial and recreational fishermen, said Hainkel, "where people's lives are at stake and people have high emotions."

After the Monday meeting led nowhere, Representative Triche resurrected a bill of his that had simply defined unattended nets, and inserted all of the language from HB 919 with just enough concessions to get by Edwards.

GCCA's Senators Hainkel and Cravins thought they had a winner when they came before the Senate Wednesday, June 14, with Triche's revamped HB 815. But the bill hung up over a one-eighth-inch difference in the minimum mesh size of trout and mullet nets.

The showdown had been forced initially because GCCA's senators had slipped wording into HB 919 that called for a four-inch minimum, which was impractical in most of Louisiana's waters. In the new bill, to appease the governor, they reduced the minimum to $3\frac{5}{8}$ inches. But the minimum had been $3\frac{1}{2}$ inches for years. The proposed increase served no practical purpose except to force family fishermen

IN THE LEGISLATURE

GCCA's Rep. Warren Triche, Sen. John Hainkel, and Rep. Sam Theriot huddle on House floor during debate on Triche's HB 815. New Orleans' senior senator, Hainkel would later receive GCCA's 1995 "Conservationist of the Year" award for his efforts in passing the "Louisiana Marine Resources Conservation Act of 1995."
(John H. Williams/*The Advocate*)

to spend thousands of dollars on new nets that they could only use for two short seasons.

Senators Nunez and Chabert offered an amendment to reduce the minimum to 3½ inches, and their colleagues lined up at the podium to express their support.

"It seems unconscionable to me that we have to ram it [a 3⅝-inch mesh] down the throats of these people because a few people, including Jeff Angers over there, want to win at absolutely all costs," Senator Bankston told the Senate. "What concerns me is that the position that the GCCA was asking me to take was not only unpalatable, but wrong," said Bankston, who further scolded Angers for "insulting members of this Senate, insulting people for no good reason."

Senator Gregory Barro, from Shreveport, allowed that he'd always voted for the GCCA's amendments, but this one was too punitive.

"Enough is enough. There's got to come a time when we balance the interests of recreational fishermen against the livelihoods of people we're putting out of business."

The irony of arguing over a one-eighth-inch difference in mesh size in the so-called "indiscriminate" gill nets may have been lost on most of the senators, but they voted 20-17 in favor of the amendment to reduce the minimum mesh size. Modest as it was, the accommodation to family fishermen infuriated Hainkel. "I wash my hands of it," he snapped, and left in a snit, trailed by Cravins.

Without their support, HB 815 went nowhere.

On Friday, June 16, three days before the close of the session, GCCA's Cravins decided to go it alone.

After Governor Edwards told the senator that he'd sign a bill with a 3½-inch mesh size, he moved for its passage. Hainkel, after noting that he and GCCA's executive director still opposed such a net, then softened his opposition.

Since mullet were easier to catch in the dark, coastal Senators Nunez and Chabert tried to strip the amendment that prohibited fishing at night. They also tried to liberalize the income provisions that fishermen had to meet to get a license. Under the provisions of the bill, said Nunez, netters couldn't make enough money to qualify

IN THE LEGISLATURE

for a license. They were unsuccessful.

The Senate approved the bill by a 33-5 vote, and sent it to the House.

When the House met on Saturday, June 17, industry supporters claimed that the bill went too far, and tried to block its passage.

"What's so different about the guy that wears white shrimp boots, speaks broken English and wants to work 14 hours a day to provide for his family? Why should we put him in a corner where he cannot earn a living," asked Bobby Bergeron. The latest version of HB 815 was "like asking a carpenter to build a house without a hammer."

"This isn't just about the white-booters in Bobby's district," countered GCCA's Triche. "It's about everybody else coming over here to rape this state."

The House approved Triche's bill by a vote of 85-17, and sent it to the governor.

GOVERNOR EDWARDS

With his last legislative session behind him, Governor Edwards waxed nostalgically at a June 20 press conference about his years at the state's helm. He also bemoaned the selfishness and lack of cooperation exhibited in the past session. "I hope in future times there will be more willingness for legislators to understand the problems of other legislators and their constituents because this state is very diverse," said Edwards, who pointed to the gill-net controversy as a classic example of "how easy it is for people to overlook something that is so important to a group of our citizens or to a geographical area and just go with what is important to them without regard to the consequences."

Edwards hoped to be remembered kindly, and more than anything else he hoped to be remembered for working to protect the interests of the general population instead of "the rifled interest of a small group of people."

"One of the greatest responsibilities and privileges of the majority is to be concerned about the minority," said Edwards.

When asked whether he'd sign GCCA's net-ban bill, the governor responded, "I don't know." A typographical error was holding him up.

The final draft of the bill said, "The department shall not issue any [license] tag to a person who does have a Social Security number." Clearly, the bill was meant to read that the department couldn't issue a license to someone who did *not* have a Social Security number but, once a bill leaves the Legislature, its wording can't be changed.

Edwards said he couldn't sign a bill if it prohibited Wildlife and Fisheries from issuing licenses to people with Social Security numbers because everybody had a Social Security number.

Lawmakers had tried to correct the problem hours before the close of the session. The corrected version passed the House but didn't make it through the Senate in time.

Edwards described the language in the bill as "obviously a typographical error, or some kind of miscommunication," and asked an assistant to look into a solution. "I certainly hope we can work some way around the problem," Edwards said.

The governor, who'd urged the participants to pass a bill less harmful to commercial fishermen, said HB 815 "represents what I thought was the best possible compromise to come out of the Legislature."

Fishermen held out the thin hope that a strict interpretation of

IN THE LEGISLATURE

legislative protocol would force the bill to stand, as is. The governor would then have no choice but to veto the bill and, with the session over, that would be that.

Could such a thing happen? Was somebody looking out for them, after all? As it turned out, no.

After the House's legal staff said the glitch was an obvious error that could be handled administratively, the governor decided to sign the measure.

Edwards had planned to sign the bill at a scheduled bill-signing ceremony on Monday, June 26, but he did not. Instead, he signed only one bill, and declined to take questions about other legislation. The governor was holding off on the net ban, said an assistant, but would be discussing it at a news conference on Wednesday, June 28.

Fishermen begged Edwards to veto the bill. In a letter she sent him on Monday, Tracy Kuhns, the director of the Lafitte-based Gulf Coast Commercial Fishermen's Association, said the bill "will cause severe hardship on the poor, hard-working people you have so avidly supported over the years. It will destroy small businesses and independent commercial fishermen. Is this the legacy you want to leave behind?" Kuhns supplemented her letter with several others from the children of commercial fishermen, also asking Edwards to veto the bill.

Joe Herring, a professional biologist with over 45 years of experience, had been appointed by Edwards to lead the state's Department of Wildlife and Fisheries. In that capacity, Herring sent Edwards a seven-page letter, asking him to veto the bill.

Senate President Sammy Nunez tried to convince him to veto the bill. If he wouldn't, at least allow the bill to become law without his signature, Nunez requested.

LET THE GOOD TIMES ROLL

State laws took effect immediately upon the governor's signature, but if an unsigned bill were allowed to become law, it didn't take effect until August 15. The delay would allow fishermen to work a little longer, said Nunez.

Edwards complied with the senator's wishes and announced at his press conference that he would allow the bill to become law without his signature.

The governor left little doubt that he didn't like Triche's bill, but said he would honor a promise that he'd made to GCCA supporters during the final days of the legislative session not to veto the "compromise" measure.

"I cannot and will not be a part of putting 6,900 people out of work, which would have a ripple effect on our economy. I am not going to sign it," said Edwards. "I just don't think it's fair. I don't think it's necessary. There is no biological evidence to support the ban.

"But, I am not going to veto it either. I find myself honestly caught by having made an agreement for a compromise that I am not satisfied fully is in response to that agreement. But, the proponents felt that it does and I am just not going to break my word to them by vetoing the bill, although I really wish it had not passed.

"It is bad, it is not necessary, it ought to be vetoed, but I gave my word," concluded Edwards. "This is the best I can do under the circumstances." Edwards said the GCCA and sports fishermen were

[Facing page]
On June 22, just days after Louisiana's Legislature crushed the state's finfishing industry, the New Orleans office of the Washington, D.C.-based Marine Fish Conservation Network targeted shrimpers with this newspaper advertisement. The "indiscriminate waste" associated with trawling had to be eliminated, said the group's ad, "in order to keep the fishing industry viable and to ensure you are able to catch fish in the Gulf!"

IN THE LEGISLATURE

Wasted Fish Stink!

So does the politics as usual that threatens fishing in the Gulf of Mexico.

Each year more than one billion pounds of recreationally and commercially valuable fish are killed by indiscriminate fishing.

Shrimp trawlers annually kill and waste:
- 34 million red snapper
- 5 million Spanish mackerel
- 650,000 king mackerel

We must eliminate this indiscriminate waste in order to keep the fishing industry viable and to ensure you are able to catch fish in the Gulf!

The Future of Fishing in the Gulf Depends on You!

As a member of the House committee overseeing our fisheries, last month **Representative Billy Tauzin** voted to weaken protection for fishing in the Gulf.

Representative Tauzin sponsored provisions that:
- Block progress in reducing the wasteful discard of red snapper and other important fish;
- Weaken protection for important fish habitat.

Call Representative Billy Tauzin (202/225-4031) and remind him that he has an obligation to protect all commercial and recreational fishing in the Gulf.

Call Senator John Breaux (202/224-4623), who sits on the Senate committee overseeing America's fisheries, and ask him to work with us to fight similar weakening amendments in the Senate.

Otherwise, the phrase "gone fishin'" might become "fishing is gone."

Marine Fish Conservation Network

Paid for by Marine Fish Conservation Network • c/o 523 Broadway Street • New Orleans, LA 70118 • (301) 953-9111

being selfish and ignoring the impact it would have on people who have made their living fishing for generations. "I hope the sports fishermen will take another look at what they have wrought here and have a little more understanding" if the issue surfaces again.

"The commercial fishermen have a good and just cause," he said, and by allowing them to fish for two more years, they had time to lobby lawmakers for a reprieve. "Maybe the next Legislature, after this bill goes into effect and the effects of it are seen, will have a different view. But we all know where the power is. There are thousands of sports fishermen on a crusade to do this."

GCCA's Jeff Angers called the bill "a victory for all the people of Louisiana. It will help conserve our marine resources for future generations and should ensure that our commercial and recreational fishing industries will remain healthy."

Since the sportsmen gave up nothing, the bill was no "compromise," reminded industry lobbyist Fernandez. "It was more like getting raped. We were more shabbily treated than any group I've seen in my 15, 18 years there."

"It's cultural genocide," said fisherman Pete Gerica. "You've got people who built this state being forced out of work by the rich and I don't think that's what our forefathers had in mind."

The new law was "the greatest example of man's injustice to man that has ever come out of this Legislature," said Representative Odinet. And Senator Nunez predicted, "Two years from now, we'll be here fighting the same battle," when the last netting season comes to an end.

CHAPTER SEVEN

NOT OVER YET

Louisiana's net-ban law was to take effect on August 15, 1995, but the Louisiana Seafood Management Council filed suit in Baton Rouge on August 11. District Judge Janice Clark set a hearing date for August 31, and issued a court order that barred enforcement of the law until then.

Encouraged by the delay, about 300 fishermen attended an August 16 fundraiser in Chalmette. Speakers at the rally thanked the legislators who'd fought for them in the past session, and cheered the Management Council's attempt to stop the ban before it took effect.

George Barisich, president of the United Commercial Fishermen's Association, told the fishermen that net-ban opponents were collecting money across the state to finance the court challenge, and asked the crowd to purchase T-shirts and bumper stickers, and to each donate $100. Barisich, a shrimper and oysterman from Chalmette, had dropped out of law school after pledging to his dying father that he would fight to save the state's fishing industry. He'd founded the UCFA, and with a core following of St. Bernard Parish fishermen had quickly built it into the state's most prominent commercial fishing group.

Chef Frank Brigtsen, of Brigtsen's restaurant in Uptown New

Orleans, also spoke at the meeting. Brigtsen told the audience that fish made up 50 percent of his menu, and that he used only freshly caught product from Louisiana. "There's no substitute for these fish," he said, signaling that restaurant owners and others who depended on the fishermen's harvest would renew their support.

Also in August, sportsman Robert G. Miller of Metairie wrote a letter to the editors of the *Times-Picayune* captioned, "Enough fish for all of us." Apparently unconcerned about advancing into the ranks of the state's angling elite, he noted that "a very unfair law was passed recently by our Louisiana Legislature—banning gill nets.

> "I have been a sport fisherman for many years and have many friends and associates who are sport fishermen. It is about time that we should stand up for the commercial fishermen. These hardy citizens of Louisiana deserve the right to earn a living from our abundant waters.
>
> "The industry could and should be regulated to ensure preservation of our natural resources—resources that belong to all of us, not only those who are sport fishermen. To stop these commercial fishermen from earning a living and supplying fish for the people of Louisiana that do not fish is an injustice to us all.
>
> "Another point that must be made is that the Gulf Coast Conservation Association is not the voice of all or even the majority of sport fishermen in Louisiana. This organization may be pitted against the commercial fishermen, but I, along with many of my sport fishermen friends, feel that there are enough fish for all Louisianians."

On August 16, the day after the net ban was originally scheduled to take effect, the FBI raided the home of Senator Larry Bankston. The feds had been tapping Bankston's phones since 1994, and learned that he'd been taking bribes from gambling interests.

NOT OVER YET

In 1991, in another of its "economic development" schemes, the Louisiana Legislature had legalized riverboat gambling and video poker. Video poker was highly addictive—a Las Vegas psychologist called it the "crack cocaine of gambling." Polls showed that 60 to 70 percent of the public would vote to ban it if given the chance. The polls also indicated that 90 to 95 percent of the public wanted the right to vote on video poker and other forms of gambling, which the politicians had not allowed.

During the 1995 session, video-poker opponents filed several bills to do away with the machines. After their arrival in the Senate Judiciary Committee, all those bills died. Senator Bankston presided over the committee.

When the Legislature legalized video poker in restaurants and bars, a Slidell legislator had inserted language that legalized the machines in truck stops as well. In December, 1994, Bankston assured a Slidell truck-stop operator that he wouldn't allow any threatening legislation to pass. On May 19, 1995, he told a Baton Rouge lobbyist that he and his friends were willing to take a blood oath to do "whatever it takes" to stop the push for a local vote on video poker.

The day after the session ended, Bankston agreed to accept a hidden interest in a Slidell truck-stop company and $100,000 in stock in a medical-waste company owned by the same truck-stop operator.

Instead of returning for another term in the Senate, or advancing to the Governor's mansion, GCCA's former champion was disbarred and sent to prison.

When it became law, the net-ban bill became Act 1316. Clearly persecutory in tone, the 25-page act contained so many hastily crafted but crushing qualifications, punishments, and provisions that time alone could reveal their true impacts. After the mullet season opened in October, two immediately proved crippling.

"Mullet are 'net smart,'" explained fisherman Daniel Hutchinson. "How many times have they run through that net before they're big enough to catch?" On a clear day, a fisherman could run his net around a school of the skittish prey species, and watch virtually every one leap back out over the corkline. Fishermen therefore worked roe mullet at night, when the fish were more likely to hit their nets.

In addition to banning night fishing, Act 1316 prevented fishermen from working on weekends. "Weekend closures" can be useful in segregating the two user groups in heavily used areas, but a blanket closure that took in all of the state's remote coastal waters was hard to swallow. Fairness aside, it was killing them economically.

Fall and winter weather on the Gulf Coast is marked by predictable cycles. As cold fronts approach, the north wind howls for a couple days—no fishing. After the front blows through, a couple of calm days allow strike netters to work. Then the winds pick up from the south, and it blows for several more days, until the cycle is repeated with the next cold front.

Under the new law, if the few workable days fell on the weekend, fishermen were out of luck. Dockside buyers also took a hit, as did their workers who unloaded boats, packed the fish, and drove the delivery trucks.

By the second weekend, they'd had enough. On Saturday, October 28, about 200 frustrated St. Bernard fishermen and supporters blocked the roads that led to Hopedale and Delacroix Island, two angling hot spots. "If we can't fish, neither can you."

The showdown began at about 5 AM, and lasted three hours. No one was arrested or hurt.

NOT OVER YET

On Sunday, the fishermen returned, and gathered first in a dark gas station parking lot for a 4 AM prayer. Representative Kenny Odinet attended the prayer, and helped maintain order. At dawn, the fishermen and their supporters lined one side of the highway, while state police and sheriff's deputies lined the other.

As sport fishermen approached, hauling boats on trailers, police officers told them that the road was open, but the private boat launches were unstaffed. Most anglers turned around, but a few drove past the shouting crowd.

When a New Orleans sportsman drove to his regular launch in Hopedale, a caravan of fishermen followed him and blocked the launch with their trucks. As he drove away, he told the police, "I think this is ridiculous. Who owns these waters, anyway?"

Claiming that they had nothing better to do, the fishermen turned out the following weekend. Their protests, however, soon became counterproductive.

GCCA executive director Angers told a reporter that 23 people had called him after the first protest, asking him to join his group. "We don't want to build up our membership this way," said Angers, but he immediately began to recruit members with the following newspaper ad:

> "Tired of having your rights violated by a small, but well-organized, special interest group? Do you wonder if anybody is working to protect the rights of the 500,000 Louisiana fishermen, women and children who believe that it's wrong to intimidate, issue threats and illegally block roads and waterways? Your Gulf Coast Conservation Association is working day and night to protect the rights of recreational fishing families. Join GCCA of Louisiana so your voice can be heard above the shouting and clamor of powerful special interest groups. Join GCCA today!"

LET THE GOOD TIMES ROLL

The first of the strike-netting seasons closed on March 1, 1996. Barred from catching finfish until the following October, fishermen purchased crab traps, worked on oyster boats, or began to gear up for the shrimp season that opened in May. While they struggled, and their lawsuit worked its way through the courts, the New Orleans *Times-Picayune* printed a punishing eight-day series, in late March, called OCEANS OF TROUBLE: ARE THE WORLD'S FISHERIES DOOMED?

In the introduction to the series, the editors wrote that "It was conceived as an examination of Louisiana's vast Gulf of Mexico fisheries. But 'Oceans of Trouble: Are the World's Fisheries Doomed?' quickly became much more. We soon realized our story was global. ...The result is a definitive story of the fragile state of the world's fisheries. They are far worse off than most people imagine, and the time to make them better is running short."

The titles of the stories, along with a few excerpts, included:

"Way of life threatened along with Gulf's vast bounty"; "The ocean is under siege. Demand for seafood is skyrocketing. The world's burgeoning commercial fleet has become a high-tech beast, capable of wiping out a fish species in the blink of an eye."; "Fish could be bad for your health"; "Overfished Waters Running on Empty"; "Hitting bottom in the World's Oceans"; "Gulf on the Brink"; "Toxic wastes render fish infertile"; "Aquaculture wave of the future"; "Catfish a Farming Success Story"; "Managing to Fail"; "Trawling into Trouble"; "Cajun chef Paul Prudhomme's blackened redfish became a national craze, and in a few months what had been a relatively unimportant commercial species was fished to the brink of depletion."; "Northwest salmon fading fast"; "Illegal fishing thrives in vast Gulf"; "Old fishing ways going under"; "The red snapper collapse"; "Tainted Waters"; "Louisiana's commercial fishers lost a battle for a dwindling

NOT OVER YET

resource when they came up against a well-organized sport fishing lobby last year."

NO FISH?

Meanwhile, back in the water, 1995 had proved a banner fishing year for sportsmen, reported the Louisiana Department of Wildlife and Fisheries, in a May, 1996, news release.

While clamoring that greedy netters were harvesting the state's "dwindling" resources to the brink of disaster, the anglers themselves caught 2.4 million redfish in 1995, a million more than they'd boated in 1994, and 1.1 million more than the 1.3 million reds they'd averaged each year over the past five. The sportsmen's haul weighed a record 10 million pounds.

Sport fishermen also landed 6.9 million spotted seatrout, nearly a million more than the 6 million they'd taken in 1994. Both years were well above the five-year average of 5.1 million fish. The anglers' 1995 trout haul weighed more than 7.5 million pounds.

Anglers also smacked the black drum—over 230,000 fish, as compared to 142,000 in 1994, and a five-year annual average of 171,000 fish.

Recreational landings of sheepshead more than doubled, from 291,000 in 1994 to 647,000 in 1995, well above the five-year average of 350,000 fish.

SOMEBODY'S WATCHING YOU

In the fall of 1996, GCCA launched a program designed to help the "hard-working agents of the enforcement division by being additional eyes and ears on the water."

In a press release announcing the new alliance, the chief of the enforcement division at Wildlife and Fisheries explained, "Coastwatchers operates similarly to Neighborhood Watch programs that are in effect in communities across the United States. The main difference is that Coastwatchers volunteers usually are in boats in Louisiana coastal waters, and are trained by enforcement agents."

The agents teach Coastwatchers volunteers how to spot violations, which they surreptitiously report with their cell phones. "The more anonymous our volunteers remain, the more effective they will be," said the chief. "The guys fishing in a nearby boat could be Coastwatchers."

CHEFS RALLY FOR REFORM

After the first netting season closed, the Louisiana Restaurant Association conducted an informal survey of its members. The restaurateurs responded that the local fish they could still get had increased 20 to 30 percent in price; the other fish they were buying, whether farm-raised or wild-caught, were imported. "We now have a lot of tilapia from Honduras, Costa Rica or Indonesia, amberjack from Florida, speckled trout and redfish from Mississippi or Mexico, and snapper from Panama or Guatemala," said an LRA spokesman.

The best cooks know that a dish can only be as good as its main ingredients, and when their source of local wild-caught fish dried up, the chefs in New Orleans decided to do something about it. Chefs Paul Prudhomme and Frank Brigtsen led the effort.

Paul Prudhomme learned to appreciate the importance of fresh ingredients while growing up on a farm in southwest Louisiana: "Whatever we wanted to eat, we harvested on a daily basis." After several years at Ella Brennan's Commander's Palace, in the Garden

NOT OVER YET

Fisherman and industry leader George Barisich, and Chefs Paul Prudhomme and Frank Brigtsen at November 17, 1996 benefit. In response to the chefs' support of the state's family fishermen, sportsmen called for boycotts of their restaurants. *(Susie Harris)*

District, Chef Paul opened his K-Paul's Louisiana Kitchen in the French Quarter.

Frank Brigtsen, an intense sport fisherman himself, began his culinary career in 1973 as an apprentice at Commander's Palace, under Prudhomme, then spent 7 years at K-Paul's. In 1986, he opened Brigtsen's in a restored cottage in Uptown New Orleans. *Food and Wine* named Brigtsen one of the Top 10 New Chefs in America in

LET THE GOOD TIMES ROLL

1988, and the James Beard Foundation would name him the Best Chef in the Southeast Region in 1998.

In a guest editorial in the Louisiana Restaurant Association's magazine, Chef Paul lamented the destruction of the fishermen's culture and way of life, and implored his fellow chefs to help amend the newly passed legislation:

> "...Louisiana is the only state, and New Orleans is the only city, where restaurants can be supplied with shellfish and finfish caught that very morning from a network of family-owned and operated small fishing vessels that go out at 1 or 2 A.M. and are back by 10 A.M. to sell their catch. My restaurant and many others in New Orleans are supplied this way, and we pride ourselves on this freshest possible seafood. If you don't think that's important, look around the rest of the United States and see how few famous dishes developed elsewhere....I implore you...to work to amend this law. We can learn to respect, conserve, and manage our natural resources. We can keep our culture, our jobs, and our tourism if we intelligently harvest our seafood....Let's work together to amend the law to allow strike nets, and to practice good conservation with intelligent quotas and seasons based on scientific data. Remember that we own this state. We own the land and waters and their resources. We elect officials only to manage these things for us, not to take them away from us."

On Sunday November 17, 1996, the Louisiana Chefs for Louisiana Seafood presented "A Grand Tasting of Louisiana Seafood."

Held in a ballroom of the Sheraton Hotel, in downtown New Orleans, the benefit featured dishes prepared by over 30 chefs from restaurants that included the G&E Courtyard Grill, Bayona, Commander's Palace, Straya, Broussard's, Bella Luna, Patout's, Pascal's

NOT OVER YET

Manale, Mother's, Acme Oyster House, Chez Daniel, La Provence and, of course, K-Paul's and Brigtsen's.

Donations of $10.00 were accepted at the door, and guests received an educational brochure that explained, "The purpose of this benefit is to showcase the superb quality and variety of Louisiana seafood, and to encourage our legislators to create balanced legislation to protect our marine resources for both commercial and sport fishermen. All proceeds from this event will be used to encourage new legislation and to help protect our priceless seafood resources for everyone to enjoy."

Supporters lined up for blocks to attend the event.

NEW REGIME

Since the 1996 legislative session was for fiscal matters only, the 1997 session offered the new alliance of chefs and fishermen its first opportunity to try to soften Act 1316. Thanks to the statewide elections that followed the 1995 session, there were plenty of new faces in the Legislature, and a new governor.

Though fishermen had been underwhelmed by the efforts of Governor Edwards in the 1995 battle, they didn't realize how good they'd had it until sportsman Murphy "Mike" Foster took office.

A friend and hunting companion of Texas Governor George W. Bush, Foster followed in the footsteps of his grandfather, who'd once governed the state. During the net-ban fight, the portly patrician was a Republican state senator from the cane country of St. Mary Parish, where he lived in a plantation home said to be more spacious than the Governor's Mansion. He'd also inherited his father's construction business, and financed his own campaign against Cleo Fields, an owlish, slightly built African-American who stood no chance of winning.

In his home-spun ads, Foster welded, drove a tractor, and assured his following that he'd placed nowhere near the top of his class at LSU. He also ran ads that decried the use of nets by commercial fishermen, and "Sportsmen for Foster" bumper stickers proliferated.

Upon his election, Foster appointed former GCCA president and Baton Rouge contractor Jimmy Jenkins to head the Louisiana Department of Wildlife and Fisheries, over the objections of many in the state who felt the position should be occupied by someone who knew something about biology.

Since Foster liked to ride his motorcycle without a helmet, his first order of business was to repeal the law requiring helmets. The "Sportsman's Candidate" also appeased his gun-owning constituency by legalizing the carrying of concealed weapons. As for his position on fishing, the chefs soon got a taste.

Before the passage of Act 1316, the state issued 1,144 saltwater gill-net licenses. Nonresidents purchased 127, which they used mostly for fishing roe mullet. The remaining 1,017 resident licenses were used for mullet as well as any other food fish except red drum.

After the law took effect, the number of licenses fell nearly 50 percent. The state issued 595 permits to net mullet, and 16 to net pompano in the "permanent" seasons. For the two phase-out seasons, the state issued 186 speckled trout permits and 149 permits for "restricted species" such as black drum and sheepshead. One person typically held multiple permits.

The second—and last—netting season for trout and all other species except mullet and pompano ended at sundown, February 28, 1997. Afterward, chefs Prudhomme and Brigtsen met with Governor

NOT OVER YET

Foster to discuss the possibility of amending the law in the upcoming session to allow strike netting to continue, and to allow the handful of the state's licensed commercial rod-and-reel fishermen to catch some redfish. The meeting was also attended by Henry Mouton, a supporter and fishing companion of the governor as well as a founder and past president of GCCA.

"No nets and no redfish" was the governor's response. And so it remained through both his first and second terms.

Commercial fishermen and their culinary supporters did their best to keep the issue alive through the next eight years. They entered bills to allow strike-net fishing for black drum and other underutilized species and night fishing for mullet, and to liberalize the qualifications for obtaining netting and rod-and-reel licenses. They also tried to make biological data equal to social and economic considerations in managing a fishery, and to establish a special fish-trawl season outside the shrimp-trawl seasons. And as Wildlife and Fisheries biologists continued to report the abundance of redfish, fishermen also tried repeatedly to regain access to that wintertime staple, using rods and reels or trotlines.

Under Foster's persistent threat of a veto, they met with little success beyond allowing some formerly disqualified fishermen to enter the fishery, and the creation of an apprentice license which did permit new recruits to enter the limited fisheries.

MILESTONES

In 1997, Texas-based GCCA named "Louisiana's best-known outdoorsman," Governor Mike Foster, "Conservationist of the Year," for his "lifetime commitment to responsible use of our state's renewable resources."

LET THE GOOD TIMES ROLL

In November, 1998, a federal grand jury indicted former Governor Edwin Edwards for extorting money from applicants for riverboat casino licenses during and after his last term in office. He was later convicted and sentenced to ten years in federal prison.

Also in 1998, GCCA gave contractor and Wildlife and Fisheries Secretary Jimmy Jenkins its "Conservationist of the Year" award "for his commitment to marine conservation."

During the 1999 legislative session, the Senate unanimously passed a resolution by Senator John Hainkel to assemble a "strategic plan" to help keep oil industry jobs in Louisiana. Senate Concurrent Resolution 40 called for the creation of a task force to talk with oil interests to keep them from moving out of the state. Hainkel said the New Orleans area had recently been hit hard after several oil companies announced that they were closing some offices and relocating them to other parts of the country.

In the same session, Senator Lynn Dean, a maverick entrepreneur who replaced seafood industry supporter Sammy Nunez, introduced a bill to let mullet fishermen work at night. Arguing against the bill, Senator Hainkel said, "A net doesn't know if it's caught a mullet, a redfish or a speckled trout. This bill allows night fishing and there is no conceivable way we can enforce it with the thin line of wildlife agents we have."

Dean said he could not believe the arguments of the opponents. "We have night shrimp fishing. You can't hide the fish. The agents can wait at the docks and inspect the fish when the boats come in. We're not talking about reds and trout. We're talking about poor mullet fishermen trying to make a living."

Three senators voted for Dean's measure.

"The senators remembered six years ago when the hottest issue in

NOT OVER YET

the Legislature was the gill net, which was used to catch redfish and trout," wrote an Associated Press reporter, in a May 5 account of the night-fishing debate. "The gill net was banned, the commercial fishing of reds was banned, and only a small allotment of speckled trout was allowed. At the time, the populations of the fish were dwindling because gill nets of up to 2,000 yards long were stretched across bays, catching reds and specks by the ton."

Hainkel told the reporter: "Reds and trout are back now and people are flying in from all over the country to fish and spend money."

WHAT COMPENSATION?

To further protect those fish from the locals, state officials weaseled out of paying the net fishermen for the loss of their livelihoods, and slipped their money to the enforcement division.

Established with great fanfare during the net-ban fight, the "Commercial Fisherman's Economic Assistance Fund" was intended to reimburse netters for the loss of their capital investments, and to provide job training. The act also provided for a portion of the money to be used by the Louisiana Department of Wildlife and Fisheries Enforcement Division. Money for the fund came from recreational fishermen who paid an extra $3.00 for their saltwater licenses from the date the act went into effect in 1995, through June 30, 1998. Sales of the "Marine Conservation Stamp" exceeded $2 million.

Of the 1,017 Louisiana residents who held gill-net licenses when the 1995 law was passed, only ten managed to navigate the maze of deadlines, stipulations, and qualifications required for compensation before the state hurriedly dismantled the program. They shared $62,034.

To receive economic assistance for approved training programs and for taking courses at state schools, fishermen had to jump through the same hoops required to receive compensation for their now-worthless gear. They also had to forfeit forever any commercial rod-and-reel or net licenses.

Not a single Louisiana fisherman elected to be "re-educated."

The Legislature used the remaining $2,012,084 to create the "Saltwater Fishery Enforcement Fund," to boost the enforcement of fishery laws on the rapidly eroding coast.

BLIND JUSTICE

Nearly six years of legal wrangling over the net-ban law was brought to a close on April 30, 2001, when the U.S. Supreme Court refused to hear a final appeal by the commercial fishermen.

After the net-ban law passed, the Louisiana Seafood Management Council had filed two class-action suits against the act, one in state court, the other in federal court. Though fishermen sued the State of Louisiana, GCCA intervened on the state's behalf with its own attorneys.

Attorneys for the Management Council argued that the law violated the fishermen's constitutional rights by preventing them from making a living, did not conserve the fish population, and constituted a breach of the public trust by reallocating the state's fishery resources to a favored group.

State and federal courts upheld the law, ruling essentially that the State Legislature could manage Louisiana's resources any way it saw fit.

Robert Barnett, an attorney for the fishermen, referred to Act 1316 as "a mean-spirited selfish attempt by a select group of recreational

NOT OVER YET

fishermen that was unwarranted and unfounded. It's just a very unfortunate series of events that has not only decimated the lives of thousands of fishermen, packers, haulers...but also destroyed an industry that has flourished in this state for over 100 years."

REDFISH SKIRMISH

By 2004, biologists were investigating the possibility that redfish were becoming stunted because so many were crammed into their shrinking nursery. "That wouldn't be a surprise. It's fairly common knowledge that when you have more individuals competing for the same food supply, then they'll all get a little less to eat, so they won't grow quite as fast, or maybe as large," said a fishery biologist at the Department of Wildlife and Fisheries.

With the retirement of GCCA supporter Governor Foster, commercial fishing interests tested his successor. Before her election as the state's first female governor, Kathleen Blanco served as lieutenant governor. During her campaign, she'd made a strong appeal for the sportsmen's vote.

Nevertheless, Senator Butch Gautreaux, a Democrat from Morgan City, filed two bills in the 2004 Legislature to repeal the gamefish status of redfish.

In response, "all 35,000 GCCA members have been contacted about lobbying their respective lawmakers and an all-out assault is being planned," announced the group's Angers. "This bill is a giant leap in the wrong direction. Sure, redfish are doing better today than they were doing when the commercials fished them to the brink in the 1980s, but they have not fully recovered."

"Here we are after 16 years and they still don't have it," responded

LOUISIANA'S **RECREATIONAL** HARVEST OF RED DRUM
(Since Commercial Fishermen were Excluded from the Fishery)
1989–2004*

YEAR	POUNDS	NUMBER OF FISH
1989	4,379,774	1,052,081
1990	3,014,515	616,604
1991	3,999,416	872,713
1992	5,833,875	1,767,938
1993	7,424,691	1,913,832
1994	5,913,250	1,382,072
1995	10,046,848	2,449,022
1996	9,358,599	2,082,396
1997	9,481,006	1,830,537
1998	6,333,248	1,427,151
1999	7,266,304	1,763,495
2000	11,693,763	2,774,046
2001	10,526,594	2,652,282
2002	9,128,555	2,041,827
2003	10,245,642	2,143,424
2004	11,513,227	2,418,339
TOTAL:	**126,159,277** POUNDS	**29,187,759** FISH

*Commercial fishing in the state ended on January 15, 1989.

NOT OVER YET

Lake Pontchartrain fisherman Pete Gerica. "It's like President Bush looking for weapons of mass destruction. They're not going to find them."

Gautreaux's Senate Bills 206 and 853 sought a 1-million-pound redfish quota, to be caught with trotlines, which were simply long strings with baited hooks attached. His legislation also gave discretion to the secretary of Wildlife and Fisheries to close the commercial fishery if such a measure were warranted.

The senator later pulled the bills out of concern for his family and staff. In an emotional speech on the Senate floor, Gautreaux told his fellow lawmakers, "What I had hoped to accomplish was a reasonable compromise that would allow commercial fishers to round out their year with a limited season of redfish." In response, he had received more than 100 phone calls, e-mails and letters from "a certain recreational fishing organization."

"Most have been respectful, but some have been threatening and just as mean-spirited as possible," he said. "I obviously bought on to this job, and the abuse comes with the territory. But my family and those who work for me did not," said Gautreaux, who explained that his wife and legislative aide had received phone calls that they considered abusive.

Blanco kept silent on the matter. By then, it was risky for any politician to challenge the Texas-based group.

From a few thousand members in the early 1990s, GCCA boasted 35,000 by 2004. As if seining the state with a small-meshed gill net, the group continually fished for its sort of folks with recruiting billboards and a long-running fishing tournament.

Modeled after the STAR tournament in Texas, where the acronym stood for "State of Texas Annual Rodeo," Louisiana's "Statewide

LET THE GOOD TIMES ROLL

Tournament & Angler's Rodeo" was a summer-long event. During that time, if an angler caught one of 50 redfish that had been tagged and released by GCCA, he or she won one of ten fully rigged boat, motor and trailer combinations that were donated by manufacturers and dealers. Among the $500,000 in prizes, local car dealers also contributed a new sport utility vehicle as a "draw" prize.

Other sponsors of the membership drive have included Champion, Bay Hawk, and Predator Boats, Mercury Outboards, Budweiser, Shimano, Lake Charles-based US Unwired, Entergy, Shell Oil, Citgo, Baton Rouge-based Audubon Insurance, *Louisiana Sportsman*, and Condea Vista. (Presumably one of those "corporations desirous of a good image among conservationists," chemical manufacturer Condea Vista caused one of the largest chemical spills in U.S. history when a pipeline at its Lake Charles facility released up to 47 million pounds of ethylene dichloride into the environment, in 1994.)

To be eligible for the tournament's prizes, anglers paid a $10.00 entry fee. Contestants were automatically enrolled into the CCA, either as adult members or as members of CCA's youth group, "The New Tide," a program designed to "shape conservationists at a young age."

In 1991, Exxon produced "A New Tide," a video about coastal estuaries, which it distributed to all CCA state chapters, in conjunction with the New Tide program.

For their further edification, CCA kids had their own page within their parents' directly mailed CCA newsletter; the "Colonial Pipeline Kids' Page" was sponsored by the world's largest refined petroleum pipeline company.

In 2003, Colonial paid a $34 million civil penalty, then the largest in EPA history, for a succession of oil spills that occurred over 20 years in nine states, including Louisiana, Georgia, Tennessee, North Carolina and South Carolina; one in South Carolina spewed a million gallons of diesel fuel into a river, killing about 35,000 fish.

WETLANDS GOING, GOING...

In late August, 2005, a storm surge associated with Hurricane Katrina virtually erased the coastal communities from the mouth of the Mississippi River north to St. Bernard Parish, which was completely inundated. Katrina also devastated communities across the coastlines of Mississippi and Alabama.

A month later, Hurricane Rita hammered coastal communities from central Louisiana west into Texas. Geologists estimated that, in two days' time, the storms washed away 217 square miles of the state.

In 2008, Hurricanes Gustav and Ike transformed more than 100 additional square miles of marshland into open water.

Between 1932 and 2010, coastal Louisiana lost almost 1,900 square miles of wetlands—an area larger than Rhode Island.

CHAPTER EIGHT

PARTING SNAPSHOTS:
POUNDS & DOLLARS

In 1994, the last full year the Louisiana Department of Wildlife and Fisheries issued such licenses, the agency sold 489 resident and 13 non-resident trammel-net licenses, and 196 resident fish-seine licenses. At that time, these licenses were valid in either fresh or salt water so the exact number of such nets used for the harvest of marine species couldn't be determined. Most trammel nets and seines, however, were used in fresh water.

Trammel nets had been widely used—prior to the net ban fishermen purchased licenses for that gear in 63 of the state's 64 parishes. In 1994, the 10 parishes with the highest numbers were concentrated in two areas. One concentration, with a third of the total, was in a seven-parish area near the Ouachita, lower Red, and upper Atchafalaya rivers, where fishermen used the gear primarily to target freshwater buffalo fish.

The other concentration was within three coastal parishes—Plaquemines, Lafourche, and Terrebonne. Coastal Cajuns in the latter two parishes had traditionally used trammel nets to target redfish: When the Legislature declared the species a gamefish in the late 1980s sales of the freshwater/saltwater trammel-net licenses in

Lafourche and Terrebonne fell nearly 60 percent, from 168 in 1987 to 71 in 1989.

After the 1995 Legislature restricted the use of trammel nets to freshwater areas only, the total number of licenses issued statewide fell from 502 in 1994 to 409 in 1996, and 335 in 1998, suggesting that at least 100 trammel nets had been in use in saltwater areas.

The state had also issued fish-seine licenses for use in both fresh and salt waters. Unlike gill and trammel nets, seines didn't require individual fish to be manually removed from their webbing, so larger quantities could be dealt with. In freshwater areas they were used primarily for the harvest of skipjack herring and gizzard shad, two oily and bony species that were marketed as bait for crawfish, crabs and catfish. Coastal fishermen seined lower-valued edible species such as bull drum, sheepshead and rays, until the 1995 Legislature restricted use of the gear to freshwater areas only.

From the 196 freshwater/saltwater seine licenses sold in 1994, the number issued for use solely in fresh water declined to 177 in 1996 and 121 in 1998, suggesting that at least 20, and probably more, seines had been in use in the state's saltwater areas prior to the ban.

In 1994, the Louisiana Department of Wildlife and Fisheries had issued 1,022 resident and 127 non-resident saltwater gill-net licenses, which authorized the owners to harvest any coastal fish except red drum, although additional permits were required to bring in some species such as spotted seatrout.

The "Bankston/Triche Marine Resources Conservation Act of 1995" created separate runaround gill-net, or "strike-net," licenses for the harvest of just two species—mullet and pompano. The act created a new commercial rod-and-reel license which allowed seafood producers to market the fish they caught with that gear. Spotted

PARTING SNAPSHOTS: POUNDS & DOLLARS

seatrout could be taken by rod and reel only, with the purchase of an additional permit.

Sales of resident and non-resident mullet strike-net licenses peaked in 1996, at 847 and 80, respectively. By 2014, the number had dwindled to 96 and 4.

Pompano strike-net license sales also peaked in 1996, at 134 and 1. In 2014, just 5 residents and one non-resident obtained licenses.

Sales of saltwater rod-and-reel licenses peaked at 48, in 1998; in 2014, resident commercial fishermen purchased 15 rod-and-reel and 16 spotted seatrout permits.

Legislation passed in 1997 led to the development of an apprentice license for the commercial harvest of saltwater finfish with mullet and pompano strike nets or saltwater rods and reels. The license required a two-year apprentice period under a currently licensed fisherman before the apprentice became eligible to purchase either the strike-net or rod-and-reel license. Individuals who had ever held a commercial fishing license were disqualified from purchase of an apprentice license.

The scant remaining fisheries apparently held little appeal for budding finfishermen: In 2000, the first year that the licenses became available, LDWF issued two to residents and one to a nonresident.

Wild-caught finfish naturally became more difficult to obtain: From 1994—the year before Louisiana enacted its net ban—until the end of 2014—the most recent year for which harvest data were available—landings of the state's 9 more traditional species declined by nearly 75 percent, from 22.7 million pounds to 6.2 million.

Black drum and sheepshead became the two most commonly available inshore species and were harvested primarily with trawls and trotlines.

From 3.3 million pounds in 1994, sheepshead landings fell by nearly 70 percent, to just over 1 million pounds in 2014.

Black drum landings never did grow to the roughly 7.75 million pounds attainable under the state's management plan, but demand for the popular item helped sustain production near the mid-1990s level: fishermen brought in 3.7 million pounds in 1994 and 3.3 million pounds in 2014.

The only other fish that didn't exhibit a marked decline in landings since the net ban was the alligator gar, in part because it was most abundant in inland fresh waters where the use of gill nets, trammel nets and seines was still permitted. (The department issued 677 resident freshwater gill-net licenses in 2014.)

A primitive animal, reaching more than 6 feet in length, the gar was armored by flinty scales. Fishermen removed that covering to expose a large slab of meat resembling the tail of an alligator; it was usually blanched, flaked and blended with other ingredients to make "gar balls" or *"boulettes,"* similar to crab cakes. Steady demand helped keep production up: Landings, which included fish caught in both inland and coastal waters—where fishermen harvested the species primarily by suspending baited hooks from plastic jugs—reached nearly 569,000 pounds in 1994, then averaged a little better than half a million pounds each succeeding year, and in 2014 totaled 506,481 pounds.

Landings of the ever-popular southern flounder fell nearly 95 percent, from just under 1 million pounds in 1994 to about 66,000 pounds in 2014. King whiting became even more difficult to find.

A favorite fish of coastal Cajuns, who called it *"robal,"* the sweet and comparatively inexpensive whiting had been widely marketed in New Orleans and across the Gulf Coast as "ground mullet." Landings of

PARTING SNAPSHOTS: POUNDS & DOLLARS

the species fell almost 100 percent, from nearly 460,000 pounds in 1994 to just 604 pounds in 2014.

Landings of spotted seatrout—the culinary favorite throughout South Louisiana and much of the Gulf Coast—exhibited a similar decline. With their nets, commercial fishermen in 1994 just met their 1-million-pound quota; after the Louisiana Legislature forced them to use rods and reels their landings fell until apparently stabilizing at around 1,000 pounds a year. (Table 1 in Appendix lists commercial and recreational trout landings since 1990.)

Commercial redfish landings, of course, remained at zero.

GCCA's 1995 legislation allowed just two species to be taken in gill nets—striped mullet and pompano.

The sumptuous Florida pompano was one of the priciest of fish, earning fishermen up to about $3.50 per pound. Though only a handful of strike netters had pursued the species, year round, on sandy beaches and bars across the coast, the new law restricted the fishery to a maximum of 8 participants who could work for just three months within a confined area of Breton Sound, east of the Mississippi River. Since then, pompano landings fluctuated widely from year to year, reached a low of just over 12,000 pounds in 2010, then rebounded to 40,434 pounds in 2014, still about 65 percent less than the modest 115,000 pounds reported in 1994.

The striped mullet fishery had been the most lucrative within the coastal net complex, and as Wildlife and Fisheries biologists had stated, there was ample room for expansion in many areas of the state. Instead, commercial landings of the species declined by more than 90 percent, from about 12.5 million pounds in 1994, to 1.2 million in 2014.

Dollar-wise, the dockside value of the established fishery didn't

COMMERCIAL LANDINGS OF **TRADITIONAL** COASTAL SPECIES
1994 & 2014

SPECIES	1994		2014	
	POUNDS	DOCKSIDE VALUE	POUNDS	DOCKSIDE VALUE
Alligator Gar	568,948	$ 350,987	506,481	$ 426,674
Atlantic Sheepshead	3,289,426	$ 1,007,221	1,079,783	$ 507,595
Black Drum	3,738,821	$ 2,531,907	3,329,234	$ 3,136,021
Flounder	974,689	$ 1,278,013	66,273	$ 146,154
Florida Pompano	114,646	$ 321,831	40,434	$ 164,179
King Whiting	459,872	$ 141,296	604	$ 669
Red Drum	0	$ 0	0	$ 0
Spotted Seatrout	1,023,687	$ 1,068,225	1,124	$ 4,769
Striped Mullet	12,560,261	$ 7,643,004	1,185,572	$ 893,222
TOTALS	**22,730,350**	**$14,342,484**	**6,209,505**	**$5,279,283**

decline to the same degree as the volume because the fish that were still available increased in price: While landings fell 75 percent, their value dropped 63 percent, from $14.3 million in 1994 to $5.3 million in 2014.

To estimate a commercial fishery's contribution to the economy, which included the value added to the fish as they ascended the

PARTING SNAPSHOTS: POUNDS & DOLLARS

market ladder to the consumer, as well as the expenditures made by fishermen, fish houses and other industry participants as they conducted their businesses, economists used multipliers.

A 1997 economic study by professional resource economists at Southwick Associates, commissioned by the Louisiana Department of Wildlife and Fisheries, found that the state's 1996 marine finfish landings of $38.8 million (which included offshore species such as tuna and swordfish) generated $256.5 million in *sales* at the processing, wholesale, retail and restaurant levels, with a *total economic effect* on the state's economy of $341.9 million. (The same study reported that marine recreational anglers spent $450.3 million on their pursuit, with a total economic effect of $944 million.)

Although the economists used a more complicated approach to develop their sales and economic-effect totals, those totals are equivalent to, respectively, 6.6 and 8.8 times the entire finfish fishery's dockside value. So applying the same multipliers to a specific fishery shouldn't be wildly off the mark.

The $14.3 million net fishermen received for their catch in 1994 therefore generated about $94.7 million in *sales,* and a *total economic effect* on the economy of $126.2 million that year. By 2014, dockside value of the 9 traditional finfish had declined 63 percent, as had annual total sales and economic effect to, respectively, $5.8 million, $35 million, and $46.6 million.

As for the leading 11 coastal species that were considered to be underutilized, and which might have grown 50-fold to support a fishery of at least 25 million pounds, their landings instead plummeted nearly 96 percent, from just under 468,000 pounds in 1994, to 20,000 pounds in 2014.

Atlantic croaker topped the list at 14,163 pounds, followed by

COMMERCIAL LANDINGS OF **UNDERUTILIZED** COASTAL SPECIES
1994 & 2014

SPECIES	1994		2014	
	POUNDS	DOCKSIDE VALUE	POUNDS	DOCKSIDE VALUE
Atlantic Croaker	73,780	$ 42,480	14,163	$ 16,079
Blue Runner	22,148	$ 7,381	430	$ 346
Bluefish	11,123	$ 3,709	0	$ 0
Sea Catfish	4,784	$ 615	233	$ 155
Crevalle Jack	41	$ 11	0	$ 0
Ladyfish	0	$ 0	0	$ 0
Pinfish	500	$ 225	0	$ 0
Rays	4,605	$ 981	0	$ 0
Sand (White) Seatrout	235,847	$ 154,488	498	$ 499
Spanish Mackerel	82,796	$ 39,650	4,993	$ 3,588
Spot	32,079	$ 8,723	0	$ 0
TOTALS	**467,703**	**$ 258,263**	**20,317**	**$20,667**

Spanish mackerel (4,993 pounds), white trout (498 pounds), blue runner (430 pounds), and sea catfish (233 pounds). Insufficient quantities of the other six species were brought in to warrant reporting.

PARTING SNAPSHOTS: POUNDS & DOLLARS

Dockside value of the underutilized fishery fell 92 percent, from about $260,000 to $20,000.

As the amount of fish harvested by commercial fishermen declined, the number caught by recreational fishermen increased.

Comparatively mild winters and a blend of other favorable environmental factors—not the least of which was the increasing area of available habitat along South Louisiana's deteriorating coastline—all aided in keeping fishery populations sound.

While commercial-fishing-obsessed sportsmen would faithfully attribute the increase to the absence of the dreaded nets, bitter commercial fishermen would just as naturally counter that the recreational fishery was simply inhaling the fish that they could no longer catch.

Of all the possible variables leading to the increase, the one that was perhaps best documented was the growth in sport fishing.

In the same session that the Louisiana Legislature decimated the production of coastal finfish, for nutrition, it created a license that authorized sport-fishing guides to operate as businesses. From the 68 issued in 1995—the first year they were available—the number of charter-boat licenses increased more than 1,200 percent, to 825 in 2014.

Like commercial fishermen, the guides stayed on the fish from day to day, which boosted the success rates of their mostly out-of-state clients over that of the majority of private anglers who preferred to fish on their own. But whether they were flying in from California or driving to their local boat launch, Baby Boomers and their kids were pushing sport fishing into uncharted waters.

According to the National Oceanic and Atmospheric

LET THE GOOD TIMES ROLL

Administration's Marine Recreational Fishing Survey, the number of fishing trips made by anglers in Louisiana—for all saltwater species, including those found offshore—more than tripled in the three decades between 1981—the first year the agency began to collect such data—and 2013—when NOAA completed its most recent survey. Anglers made an estimated 1.4 million trips in 1981, and over 4.6 million in 2013; in 2004 sportsmen made an all-time record 5.2 million fishing trips.

Louisiana's recreational harvest of inshore fish exceeded 28 million pounds in 2013.

Landings of the lesser-targeted species, most of which were taken incidentally to the pursuit of redfish and trout, totaled 5.1 million pounds. Listed in descending order of volume, they included 1,668,090 pounds of black drum; 1,105,993 pounds of southern flounder; 1,003,450 pounds of sheepshead; 512,133 pounds of gafftopsail catfish; 374,430 pounds of sand (white) seatrout; 317,766 pounds of Atlantic croaker; 77,096 pounds of striped mullet; 65,272 pounds of Spanish mackerel; 8,550 pounds of southern kingfish (king whiting); 1,231 pounds of crevalle jack; 531 pounds of pinfish; 149 pounds of spot; 132 pounds of bluefish; and 102 pounds of Florida pompano.

As for the two premier species—spotted seatrout and red drum—sportsmen in 2013 landed nearly 9.4 million pounds of trout and an all-time redfish record of 13.6 million pounds.

Commercial fishermen had netted trout until February 28, 1997. After that date, they could use only rods and reels, if they qualified for such a license. Commercial trout landings from 1998—the first

PARTING SNAPSHOTS: POUNDS & DOLLARS

entire year rods and reels were required—through 2014 totaled 493,940 pounds.

Recreational landings of spotted seatrout from 1998 through 2013 (the most recent year such data were available), exceeded 152 million pounds, more than 300 times as many.

Louisiana's red drum gamefish law terminated the commercial harvest on January 15, 1989. In the first two weeks of that year commercial fishermen landed less than 25,000 pounds of redfish; they brought in an additional 11,500 pounds over 4 years in the 1990s, presumably in association with research on the species.

Commercial landings from 1989 through 2014 therefore totaled just over 36,000 pounds. Recreational fishermen in Louisiana, from 1989 through 2013, brought in 6,000 times that number.

Indeed, the anglers' redfish harvest of 221 million pounds during that 25-year period was more than four times the amount reported by the commercial industry since record keeping began in 1887.

• *Part Two* •

THREE LOUISIANA FISHER FOLK

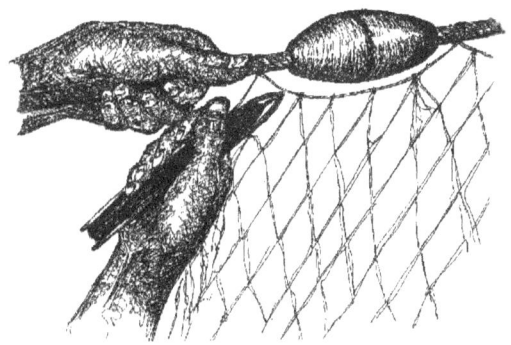

LOUISIANA FISHER FOLK
CLIFF GLOCKNER
LACOMBE

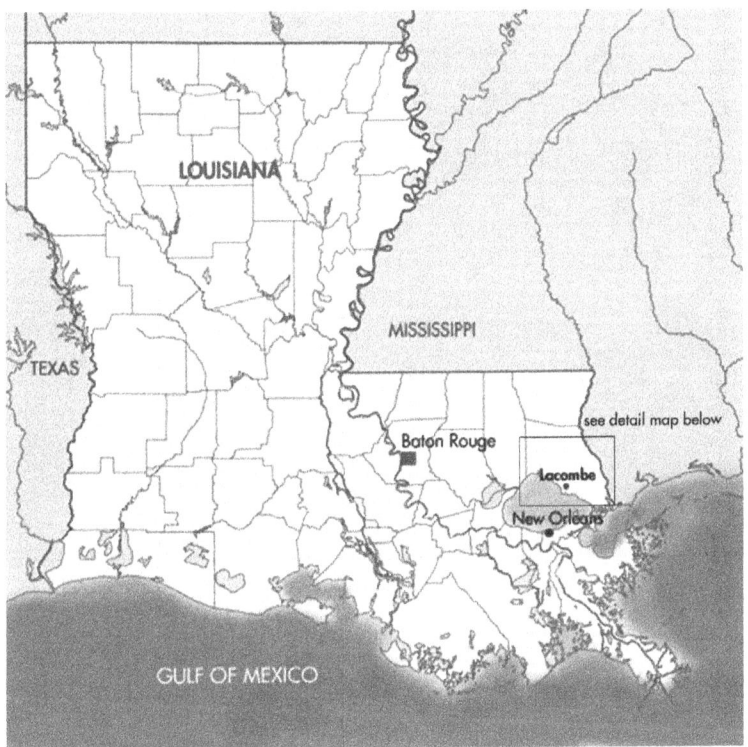

Lake Pontchartrain is the nation's second largest inland saltwater lake, but unlike the largest—Utah's Great Salt Lake—Pontchartrain is an estuary that's linked to the Gulf of Mexico. Tidal waters flow into the lake from the east, through a couple of major passes, re-seeding its waters each season with crops of blue crab, shrimp, and a variety of marine finfish.

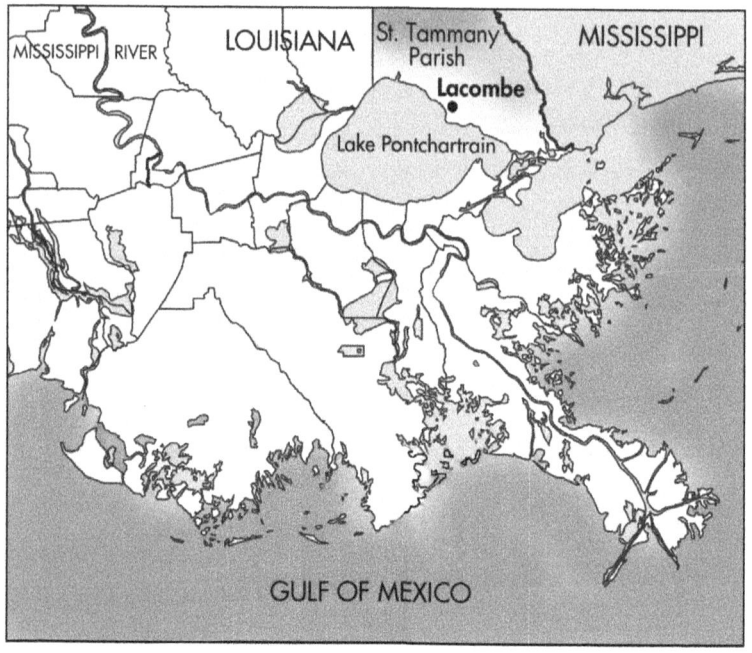

Lacombe, on the north shore of Lake Pontchartrain

"The Lake" is as far inland as you can get in Louisiana and still catch saltwater species of fish such as red drum, black drum, white trout, speckled trout, sheepshead, flounder, mullet, croaker, menhaden, crevalle jack, even giant tarpon.

Until recently—geologically speaking—the Gulf of Mexico encompassed this immense estuary. That began to change about 5,000 years ago, when runoff from melting glaciers caused the lower Mississippi River to shift its course eastward from what's now known as Bayou Lafourche, to near its present location. Within 2,000 years, the river had piled up enough sediment on the continental shelf to partition the lake from the Gulf of Mexico with the tenuous delta that currently

CLIFF GLOCKNER LACOMBE

supports Plaquemines, St. Bernard and Orleans parishes. Metropolitan New Orleans is situated on the upper edge of this delta, along the lake's southern shore.

Across the lake, its northern shoreline traces the Gulf's ancestral edge, and the land there is older, higher and covered with the piney woods that characterize the inland tiers of all three central Gulf states. Most of the lake's north shore lies within St. Tammany Parish, which had been dependent on forestry and farming until "The Causeway" was completed in the 1960s.

The world's longest bridge over water, the 24-mile-long span shaved nearly an hour off the drive between New Orleans and the North Shore, making the commute from a new suburban home a feasible and attractive alternative to life in the City of New Orleans.

St. Tammany's now one of the state's fastest growing parishes. In 1982, urban areas occupied a little over 22,000 acres, just 3.1 percent of the parish. By 2000, areas classified as urban had more than tripled to nearly 71,000 acres, nearly 10 percent of the parish. Researchers at the Coastal Research Laboratory at the University of New Orleans estimated that, by 2050, nearly a third of the parish will be covered with concrete or blacktop.

A 17,000-acre swath along the lake's shoreline will not be: The Big Branch Marsh National Wildlife Refuge protects a unique variety of ecosystems, from submerged grassbeds, sandy beaches and marshland, to pine islands, pine ridges and hardwood hummocks.

As a younger man, Lacombe fisherman Cliff Glockner shot game here for the market, and trapped furbearers. He has spent his life harvesting the marine resources that used the marshes as a nursery ground. That the refuge takes in Cliff's old stomping grounds is no coincidence—he and his wife Connie were prime movers in its founding in 1994.

"The refuge was Cliff and Connie's idea. They thought that up a long time ago," recalled Neil Armingeon, former environmental director of the Lake Pontchartrain Basin Foundation, a Causeway Commission-funded organization dedicated to preserving the environmental quality of the lake. "Cliff and Connie were fighting shell dredging, along with Claire and Jerry Crawford, another commercial fisherman. They were actually doing alright but then this study came out saying that there needed to be some organization to manage the lake. So Cliff and the others kind of hitched their star to that wagon. And actually it was better for the foundation, in a way, because the foundation was one of the groups that was credited with stopping shell dredging. But actually that whole campaign was organized by Cliff and Connie, and I don't think anyone would even argue that.

"That was back in around 1990. I got with the foundation after that, and started to work with them on the refuge.

"The Crawfords lived in Big Branch, a little community that was named for a branch of Bayou Cane. And across the street from Jerry's place there was a guy who had 24½ acres of land that he was going to turn into condos. But Tulane Law School helped them stop the coastal-use permit, and after that we decided to try to buy that land for a little park.

"We talked to the owner and he told us that he'd sell it for $250,000. Of course, we didn't have any money but Mark Davis, who was with the Coalition to Restore Coastal Louisiana, was up in D.C. and he just cold-called on the Conservation Fund. He got the name of a guy, came back, we called the guy and I think it was just one of those kinds of things that happens miraculously: We told him that we needed $250,000, and he agreed.

"And then Cliff and Connie said, 'Wait a minute.' We were sitting

CLIFF GLOCKNER LACOMBE

Cliff and Connie Glockner at their home on Bayou Lacombe.

there scrambling to buy 24½ acres and they said, 'Hey, that's nothin'. That won't mean anything.'

"They're the ones who overcame our cowardice. We were petrified at getting the money for the 24½ acres while Cliff and Connie came and unrolled this map, and we all kind of stared at it.

"There was all this land between Bayou Cane and Highway 11. But the first meeting, we weren't smart enough to think about all that. We were just thinking about the stretch between Bayou Cane and Bayou Lacombe—literally all the land between Cliff's house and Jerry's house.

"Then U.S. Senator J. Bennett Johnston had a town meeting. We got that little map and put together a presentation. One of his aides told us, 'Y'all have got ten minutes to make a pitch to the senator.'

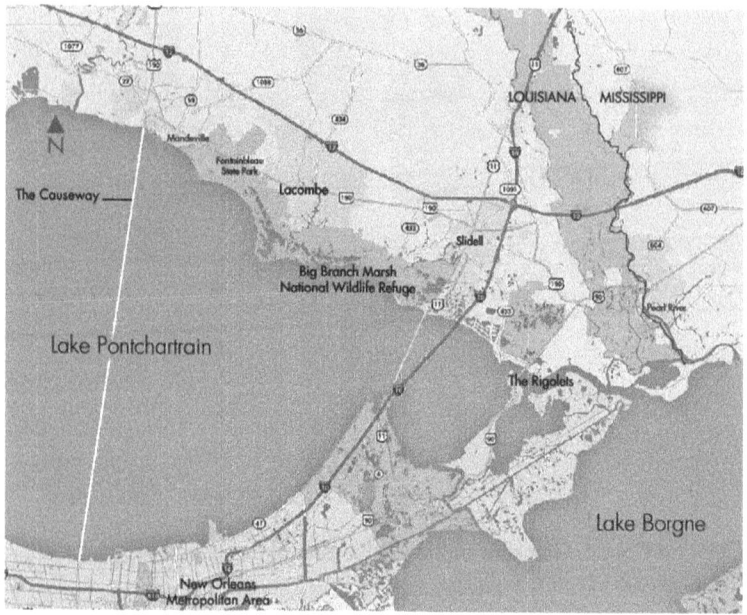

Location of Big Branch Marsh National Wildlife Refuge on north shore of Lake Pontchartrain.

And I had no idea what ten minutes of J. Bennett Johnston's time was worth. We were just naive as hell.

"We unrolled the map and pointed out, 'Here's this, here's that, and this is what we want to do.' And old Johnston, he looked up and said, 'I like that.' And he told his aide, 'I want you to help these people.'

"And that was it. J. Bennett was chairman of the Senate Appropriations Committee and, all of a sudden, our stock went through the roof, baby!

"There was one single landowner who owned 3,700 acres and, at the time, there were all kinds of plans. There were plans to turn that

CLIFF GLOCKNER LACOMBE

Cypress Bayou, in Big Branch Marsh National Wildlife Refuge.
(Tom Carlisle)

whole area into nothing but wall-to-wall condos. There were more crazy-ass ideas drawn up on that land than you can shake a stick at. But we set up a meeting with the owner and told him, 'We have no money but we have a chance for your family to do something that will mean something to everybody.'

"And he told us, 'I'll tell you what I'll do, I'll give you the right of first refusal.' So we went back to the guy at the Conservation Fund and told him, 'You know what? You can forget those 24 ½ acres. We want to buy 3,700 acres.' Then Senator Johnston brought in the U.S. Fish and Wildlife Service, which started the planning process and, to the Service's credit, they were smarter than we were. They didn't stop at Bayou Lacombe, they went all the way to Highway 11.

"That's cuttin' corners, but that's pretty much what happened. Once we got the 3,700 acres, people started calling out of the woodwork. 'I heard y'all got money and I want to sell.'

"It got up to 16,000 acres while I was there and I think it's going to keep getting bigger."

In the 1960s, Cliff and his wife Connie built their home on Bayou Lacombe, which now lies in the heart of the federal refuge. It's just a short run from their dock to the lake where Connie ran her crab traps and Cliff trawled for shrimp and netted fish. Primarily targeting black drum, roe mullet, redfish and trout, Cliff worked from his 22-foot fiberglass skiff, probably the only boat in the state rigged with a net reel in the bow, in the style of West Coast salmon fishermen.

A naturalist who saw himself as just another link in the food chain, Cliff worked hard to improve the environment of those living resources that sustained his family. Besides his environmental work, Cliff worked tirelessly to protect the interests of North Shore fishermen.

During the net-ban fight, Cliff spoke for the 78 licensed net fishermen who, in 1994, had landed a million pounds of marine finfish into St. Tammany Parish. (This with redfish still off limits and with future fisheries—such as the crevalle jacks that stormed the lake each summer—completely undeveloped.) But the state's politicians in the mid-1990s weren't listening to commercial fishermen.

Before the fateful 1995 session, early morning commuters could watch fishermen lift their stab nets along the Causeway, harvesting fresh flounder, trout, drum, bull croaker and other tasty species. Afterward, the only netting allowed within Lake Pontchartrain's 600 square miles was during the mullet's short spawning run.

By 1997, the number of licensed net fishermen in St. Tammany fell

CLIFF GLOCKNER LACOMBE

to less than 30, and their landings of marine fish into the rapid developing parish totaled just 8,000 pounds, less than one percent of their pre-ban catch.

Both Cliff and Connie suffered heart attacks during the fight; when it was over they, like so many other embittered coastal fishermen, gave up on the cause and found something else to do.

"That's when I got enough of it," said Connie. "When we were spendin' more time in Baton Rouge, fightin', just so you could get out here and bust your ass to make a livin'. When you've spent more time fightin' your legislators to let you make a freakin' living, which is hard enough by itself, I'm finished. I am teetotallin' finished. You're just beatin' your head against the wall."

After the sports killed the fish business, Cliff took the net reel off his skiff, and he and Connie remodeled his net shed into Glockner's Place.

"They've got a wildly successful restaurant now," said Armingeon, in a 2000 interview. "But Cliff is no restaurant owner. He's a fisherman and when they stopped the fishing, he was never the same. A part of him died."

The Glockners' isolated home is perched high on pilings and overlooks an expanse of the wetlands that they'd helped preserve. On a steamy July day in 2000, I visited them, to get Cliff's story. After five years of not being able to fish his lake, he was not a happy camper:

We've got two kids, a boy and a girl. Thank God they didn't get in the fishing business. My girl just got out of the Air Force, she works for BellSouth. My son's 25 and works offshore, out of Port Fourchon.

Connie and I have both been here since we were just kids. Her dad fished and trapped. He had a camp on Bayou Lacombe, by Goose

Point, and they used to spend a lot of time out there.

He was from Mandeville. He had fish markets there and, later years, he started a restaurant. In fact, the restaurant's still there—"Rip's."

I was born on the beachfront at Mandeville, 61 years ago. My dad was a fisherman, my grandfather, all my aunts and uncles.

They fished for anything, according to the season. Fishing for fish was big in the wintertime—trout, redfish, sheephead, drum, all that. They haul seined and trammel netted. They did a lot of haul-seine fishin' in the lake back then.

When I started, they were haul seining for crabs too. Softshells, in the shedding beds in the grass. They had sandbars off the shoreline and in between them, it was bog, and all those crabs would get down in there in the grass when they was shedding.

What happened to that? The shoreline got screwed up with a hurricane, the spillway opened, and the shell dredging was washing it back. Any time you take a barge load of material from one place, a barge load of material has to come from someplace else. So you had shoreline erosion.

We sold our fish over here. They had fish markets over in Mandeville, all over on the North Shore. And what we couldn't sell over here, we brought to New Orleans, the French Market. That was back in the '40s and the '50s. By the time of the late '60s, all that in the French Market was gone.

We sold to Sal Piazza, Battistella. That's where they all got started, right there in those stands in the Market.

We'd sit on a stand all night. If you sold during the day you never got anything. They'd always tell you the market was flooded, and there was no price on it. But the restaurants bought from 12 o'clock

CLIFF GLOCKNER LACOMBE

on. So you'd hang in there 'til after midnight, that's when you got your top dollar.

You know, things was hard to get rid of in them days. We used to produce hundreds of thousands of dozens of crabs, softshells. We used to carry 'em to Martin Brothers in New Orleans, Katz and Bestoff, they used to buy 'em when they still had delis in their drugstores. We sold 'em to the Roosevelt Hotel, La Louisiane, all those places. Redfish, everything.

We'd sell 'em game in the winter—ducks, woodcock, rabbits. Oh, when I was a kid that French Market would be lined up with rabbits. The rabbits, way up into the '50s, you could kill all you wanted. Then they put a limit on 'em, but you could still sell 'em, and then they made you quit selling 'em.

We'd shoot 50 to 75 woodcock a day, with bird dogs. Quail, we even sold snipe. We'd make burns for the snipe to come into. We'd trap in the morning and go hunt birds in the evening. It was fun, but it'd work the shit out of you. And another thing, if you'd kill that stuff, you never had all these freezers and ice houses. You couldn't just go to the corner and buy a bag of ice. You had to do something with it, you had to get rid of it.

You hear about guys going out there every day and killing a hundred ducks? Well, if you believe that, you ain't never picked a hundred ducks, I can tell you that. You might o' shot 'em but you get 'em out o' that marsh, you bring 'em home and you pick 'em and you ain't goin' back for a while, son.

I hear all kinds of bullshit. I know a lot of people killed a hundred ducks, but what they couldn't give away, they threw away. But that ain't hunters, that's shooters. There's a lot of shooters.

LET THE GOOD TIMES ROLL

It's like fishin'. Same thing. A lot of 'em never showed up in the fishin' business till they could go set out a net.

I was one of the first ones to use a gill net around here, and I knew then it was the beginning of the end if it ever got out. And it got out. Me and Junior DeFreitas were just looking for an easier way to catch fish. We were thinking, "Why the hell won't a gill net work? They work in fresh water."

This was way back in the early '60s. I said, "Shit, let's go and get one and give it a try."

We went and got 'em from the Steve Miranovich Trawl Company in Biloxi. I bought a thousand foot and he bought a thousand. And we was striking trout and stuff. The trammel net was so aggravating. It'd catch all kinds of little shit you didn't want. So we got the gill nets and just targeted what size fish we wanted by the size mesh. And that's how it got started.

Then they started sticking 'em out there and leaving 'em. And, "Oh boy," I said, "we got bad news here."

It started with trout around here, and down the river it started with drum and redfish. They started sticking 'em on all the points. I said, "Oh Lordy, here come the merry-go-round."

When they closed it, I was fishing from a 22-footer. I had a net reel on it but since I took the reel off, it's nothing but a sport boat now.

Why a net reel? I was fishing by myself. Why bust your ass pulling on a net when you can let the reel pull it up? It was electric and I could put 8,000 feet of net on there, 1,200, 1,400 pounds of net. It kept your nets organized, you know?

I used to strike fish. I'd go all over, along the bank, out in the

CLIFF GLOCKNER LACOMBE

middle. I've already seen black drum so thick they were jumping out of the water. Or you could see 'em boiling in the water. But mostly you'd look for muddy water where they were digging in the bottom. It takes awhile for that mud to reach the top so I'd get the boat out ahead of the muddy water and put the net down there.

The net ban? They had everybody stirred up. They had a bunch around here ranting and raving. This all started back with the damned redfish. That was the root of the trouble.

Redfish has always been a rock in my shoe. I would always catch the shit out of redfish, but there wasn't a market on 'em. A dime, a nickel, 18 cents. You thought you died and went to heaven if you caught one for 18 cents a pound. If he was over 10 pounds, he was a bull. Fifty cents, I don't care if he was 20 foot long, he was still 50 cents.

We was fishing redfish, but in them days we would fish for "makeup," we'd never target one thing for long.

If you made most of your money at shrimping, you fished a few fish, dredged a few oysters, you trapped muskrats and all that. And when shrimping would come around, that's what you'd really put your effort into. Or if you made your money crabbing, you'd do the other things for makeup. And when crabbing season would start, you stuck with the crabs. You gave everything a rest.

Then they started targeting fish. Then the oil patch failed and all them morons jumped in. Then the Vietnamese—the Catholic Relief brought them over here and stuck 'em all in the fishing business. And let's not forget the Florida fishermen, they jumped in there.

So when you put that pressure on, you know...but they blame Paul Prudhomme with all that blackened seasoning. That's bullshit.

LET THE GOOD TIMES ROLL

It all got out of hand, but instead o' putting it under control, they just got rid of it. That's stupid.

But that started everything. The sports started coming out of the woodwork after that: "They're catching dolphins, killing birds," and it's like I told them at a meeting, "If I catch a goddamned duck in my net, y'all have a stroke. But you go out there in the marsh, you take a shotgun and blow 'im into a powder puff, you're a hero."

They shoot 'em up, cripple 'em up, everything go off and die, oh that's a sport. I catch a duck incidentally in my net trying to catch a fish to eat, well they all have a dying stroke, they come foaming from the mouth.

They brought that dolphin in and all that, and said they found it in a gill net. You're so full o' shit, fella. I fished all of my life and I can tell you I never caught a dolphin in my net. Never, none.

That was a set up. Look at the fish in the net, they had ducks in the net, they had redfish in the net, they had dolphin in the net. What other goddamned endangered species they had in the net? Spotted owls? Why not put a couple o' them in there and be done with it?

Where were the pogy? Where were the garfish? Where were all the other trash fish? They weren't there. No shad, nothin', just redfish and dolphin and *ducks*, Oh, give me a damned break.

Then they hauled it off and had a goddamned funeral for it. I said, "Now you really lost it!"

If we could've come up with the money, we could've bought their man. He'd have walked the other side of the fence. Just put that dollar there.

But we go give a lobbyist $10,000. You don't think that guy's gonna go do you any good? He's gonna take your money and say,

CLIFF GLOCKNER LACOMBE

"Oh well, I tried."

What's $10,000 to them people up in Baton Rouge? He ain't got no doors that he can open. You got to get somebody that knows what the hell's going on. Knows who to talk to, knows how to put those riders on other bills at the last minute, to get what you want fast. That's how you got to do it. That's how the GCCA wound up doing it.

Them sports thought they could scream their lungs out and people up there was gonna listen to them. That didn't work for years, but then they caught on. They got a little savvy. That bunch from Texas, with the oil companies, they knew how to work it. And they did too.

They wanted the fish just for sport. That's it. And so they got it. I hope they're happy.

After they took everything else, I qualified for mullet. I had all that. But the way I looked at it, what am I gonna do, jump in there and then some crazy SOB is gonna come along and say we're destroying all the mullet and close that too? You get all settled in and here comes some goofball?

I think I made a good move. I could have got a mullet license, or that sport license, that rod-and-reel license. But do you think I'm gonna go up there and have a moron tell me how to take care of fish? You think I'm gonna sit there and listen to that shit? I don't think so. That makes my blood boil just thinkin' about it. To get the rod-and-reel license, you had to pass a test to show them that you know how to care for your fish. And you had to sign a paper that you would never net-fish again. Yeah, right.

I used to fish with an old boy, Bob Freire, from down in Hopedale. He never used over 300 foot of net, and could catch a boatload of fish any time he wanted. Three hundred foot of trammel net, that's all he had, but after the net ban, he couldn't qualify for a mullet license or a rod-and-reel license because he'd never had a gill-net license. All he had was trammel nets.

We fought it as hard as we could. You'd fish and go up to Baton Rouge, fish a day and go back to Baton Rouge. It got so you'd be up there with your slicker suit on. It ran you down. I had a heart attack and so did Connie. She's had five bypasses, three of 'em worked and two didn't.

During the fight, Eddie Deano was our representative in the Legislature. He's a fine guy, but Eddie couldn't help me. He would probably have liked to. He called and asked me, "What can I do?"

I said, "Let me tell you what you do. Put your tail between your legs and hide. And don't say nothin'."

And he said, "Why?"

"Because you're gonna commit political suicide if you come help me," I told him. "Don't stick your neck out. If they ask you about it, just give them a little answer. Don't go up there and get in a fight, because you ain't gonna win."

See, they only had a handful of net fishermen over here, but it's full of sports. St. Tammany had something like 20,000 of 'em. Then you had the big politicos over here, too, senators like John Hainkel and "Sixty" Rayburn.

Hainkel's been in the Legislature since the '60s or '70s, and he's strictly against the commercials. He's got Mandeville, Madisonville, Pontchatoula—his district runs from New Orleans all the way up there.

He's well liked in the Legislature, and I like him myself—he was

CLIFF GLOCKNER LACOMBE

just over here the other day, catching undersized crabs off the dock. John did a lot of good things. I've seen him go my way on a lot of issues, but me and John Hainkel was on two totally different sides of the fence when it come to fishing issues.

With the net ban, John seen it his way and I seen it mine, but I tried my best to explain it to him.

I brought a bunch of fishermen from around the lake in his office, to talk to him, to try to convince him, and I think that was the biggest mistake I ever made in my life. They jumped on his case and one of the guys from Bucktown, he made an ass out of himself. John looked over at me and I told him, "What can I say?"

"Sixty" Rayburn, he'd been a senator over here for years and years too. I could maybe have got him to go against it, but he wasn't gonna stick his neck out either. Sixty knew what a big political issue that was. You just had so many sports move over here from New Orleans that you couldn't expect these people to try to help you.

Because we had never did nothing to help our own selves. Before, what we should have done, we should have went to every dog-and-pony show that they put on and *told* people what we did, *showed* 'em what we did, and explain how it was a benefit, you know? Every old lady's tea party you come to, garden party, you educate them. But we never did that.

They went at 'em like bulldogs. And the old-time political bulldog thing don't sell good no more. You look like trash. Now, you dress nice, you act nice and you talk nice. You catch more flies with honey than you do with vinegar.

A lot of us are rough around the edges. And people don't like that. You have to go right in the middle of all these people, whether they're environmentalists, whether they're sport fishermen, and you don't get

LET THE GOOD TIMES ROLL

mad and put your finger in their face. You don't have to stand there and take their crap but you have the answers for 'em. And that's as good as knocking the shit out of 'em, when you got all the answers and they ain't. That's how you beat 'em. You don't beat 'em with a bunch of lies, you beat 'em with the truth.

Their lies? That's gonna come back at them with a vengeance. You don't win everything in this world with lies. It comes back at you. There's that old saying, "What goes around comes around?" And it's gonna come around on them, watch what I'm telling you. The same lies that they told are gonna come back to haunt 'em one day.

A lot of fishermen couldn't understand these legislators and their local politicians. They were voting for 'em, but they really wasn't helping 'em. We were such a minority. The politicians that are in the areas where there are a lot of commercial fishermen, you know, their constituents? They're gonna vote for them. When they get up there to Baton Rouge, they're gonna vote for the fishermen.

But it don't make any difference because they don't try to solicit all these other legislators to vote their way. So when they get up there and they vote, they say, "Well, I tried to help you, but they just couldn't see it my way. We were outvoted by North Louisiana."

They didn't really go up there to help, they didn't get in any fight. Because they need things themselves, and they don't want to butt heads with these other people because when it comes to a vote for them for something, they'll vote for them. And they'll be telling their people up there the same story, "We were outvoted by South Louisiana." That's how it works.

Edwards? He didn't do nothin' for us. But you ain't gonna get nobody from South Louisiana to tell you that he was a piece of shit. Those Cajuns love Edwards. He stole from the people. He did enough when he was in power that if he would have did good with it, this state would have been number one in the country. But he did it to pad his own pocket, he sold political jobs, everything, any kind of dirty deal you can name, that bastard was in on it.

Paid off to not veto the net ban? I wouldn't doubt it, but I wouldn't accuse him of taking money for that. I would think he was smart enough a politician that he'd just sit back and let it take care of its own self. He was very slick. King Edward. If he would have taken money for that and that would have got out then that could have hurt him. I say he was smart enough to just stay away from it.

Is predation a big problem? I really don't believe in that. You hear that the black drum are eating all the oysters. And that the redfish are eating all the crabs. But I remember when the schools of redfish used to come up and the water would be like blood for blocks around the boat. And we was catching boatloads of crabs.

I heard for years, "A hardhead catfish ain't good for nothing. Why'd God make a hardhead?" Just because you can't sell it and make money, or make a sport out of it, it's no good. But he's a scavenger, he goes around and eats up your garbage.

When I was young, the sport fishermen wanted you to kill all the garfish: "They're eatin' all our bass!" Now they're $1.50 a pound, or more.

Just like those bowfin. Choupique. The sports would catch 'em and throw 'em out on the bank. Now they're making caviar from their eggs!

"No good for nothing." Man, you're walking on my wrong side

when you start talking that bullshit. There's a balance. That fish does something. You might not understand it, you may *never* understand it but there's a reason he's there.

I'm a conservationist, I believe the redfish has got a right to go eat a crab if he wants to, and I believe all the other animals out there got a right to make their living, including this animal right here!

There's so many fish right now, it's unbelievable. Just like, the outdoor writer with the *Times-Picayune* called the other day and asked me, "Why is there so many fish? Because they stopped gill-netting?"

I said, "Look, plenty of times when you see a lot of game, or a lot of fish, it ain't because of *one* thing. You don't do one thing and it's, "Oh, it's all gonna be a big panacea," you know?

I said, "We ain't had a big bad freeze since 1989. And we've been doing a lot of work in the environment and everything, cleaning up the water. You got better water quality. We ain't had a real hurricane in 32 years. It's a bunch of things, it ain't one thing."

You get a bad storm this year, the fish population's liable to go to naught, depending on how bad that storm is. A lot of damned fish wind up out there in the marsh and that's the end of them. People don't think that, but shoot, those little bitty trout, juveniles coming up? I've seen them by the millions in the marsh. All laying in the grass dead, after a storm. Shrimp? I've seen ditches full of dead shrimp.

What about natural diseases. Those sports don't realize that fish populations don't do this [gesturing a straight line], they do this [up and down]. And it's been like that since the beginning of time, whether it be crabs, shrimp or fish. Up and down.

But take the redfish ban. One year after they did it, the reds

CLIFF GLOCKNER LACOMBE

were back up in record numbers. So GCCA stands up and says, "See, it worked!"

I said, "Yeah, sure it did."

Look at this marsh, there's just enough muskrat out there for seed. You almost don't see a muskrat no more. But there ain't been no trapping in this marsh for at least ten years, and there's no buildup of muskrats.

I've come to the conclusion that the only way you can build up a big muskrat population, you have to put pressure on 'em. And that'll stimulate 'em to breed and build 'em up. Take that pressure off and they'll settle down and kick back. That's true of a lot of animals.

Fish? Put too many fishermen in there, put too much pressure and you can hurt 'em. You can put too much pressure. However, we're going the other way now. You got to strike a balance. And how you balance things out is very easy. You set you a number and you let it run for a couple of years. And then you check those numbers up. And when you see a dropoff, you set the numbers back, because you're doing something wrong. Or nature's doing something. But they don't do that, they chisel things in stone.

Right now, what would be wrong with catching ten million pounds of redfish? You couldn't put a dent in them. But no, they'll give you a million pounds or something like that. They think that's so much, but that ain't nothing. They ain't got no idea of how many fish are out there, and that's what's wrong.

Just like after the 1989 freeze, the guy that was runnin' the GCCA then was up there at the Wildlife and Fisheries building, and he says to me, "What do you think about all the millions of pounds of fish that's dead out there?"

I said, "You want to know why you see millions of pounds dead?"

LET THE GOOD TIMES ROLL

And he said, "Why?"

"Because there was *billions* there," I told him. "You think they *all* died? Well, have I got news for you. Most of those fish went offshore, and in two years they'll all be back. They'll start spawning and there'll be a helluva crop."

It come out just like I told 'im. In two years they were up to their asses in fish. But it didn't matter. That freeze was a helluva good excuse to get rid of us, wasn't it? If they were really worried, they could o' put a moratorium: No commercial fishing for two years and no recreational fishing for two years. But do you think they'd do that?

Just like this here, this used to be a "sanctuary." You couldn't fish for four miles out from the bank. We got it cut back to a mile. You ain't supposed to fish commercial. You can sport fish though. Now how is that a sanctuary? It's a private recreational area. But that's the politics in this state.

What we call the "middle ground," over there between the I-10 twin spans and the Rigolets? It was a "sanctuary" too at one time.

I was coming across the lake over there, back when that was all closed to us, and the shrimp was in there like hair on a dog's back. I told my deckhand, "We're gonna put down and nail these bastards."

We put down and was dragging and a helluva squall started coming from the west, and I mean just as black and as ugly a damned squall as you'd ever seen in your life.

We was about three quarters through that drag, and all of sudden right underneath the twin spans, here come the "conservation" boat. I told my deckhand, "Look, you don't say anything, you let me do the talking. Get your ass over there on deck and make like you're doing something."

So they come alongside, "Get it up!"

CLIFF GLOCKNER LACOMBE

And I said, "What's wrong, Cap?"

He said, "This is a sanctuary."

I said, "I don't know nothin' about no sanctuary. I didn't know you couldn't trawl here. I hear 'em on the radio, they say they're trawling by the bridge."

"That's the Causeway, you stupid sonofabitch! Where in the hell are you from?"

"Golden Meadow, Cap."

And he said, "Get it up!"

"Okay, I'll pick up."

I picked up and, boy, that net was full! But by then, that squall was right on top of us, and he was getting scared.

He said, "Look, get up and get out of here, and don't let me catch you back here. I'm gonna let you go this time." So he took off running for the Rigolets. He got almost in to the light over there and he shut that motor down. You know, it finally dawned on 'im who I was.

Two weeks later, Connie was back here in the boat, sorting some crabs, and he come in and pulled up alongside of her and said, "Boy, that old man of yours pulled a freaking slicker on me!"

She said, "Really? What'd he do?" And he told her the story and said, "When I got into the Rigolets, it dawned on me. That no good sonofabitch, that was Cliff Glockner!"

POSTSCRIPT

In August, 2005, Hurricane Katrina destroyed the Glockners' restaurant; they did not rebuild. Cliff died five years later, at age 70.

LOUISIANA FISHER FOLK

KERRY LEBAUVE
COCODRIE

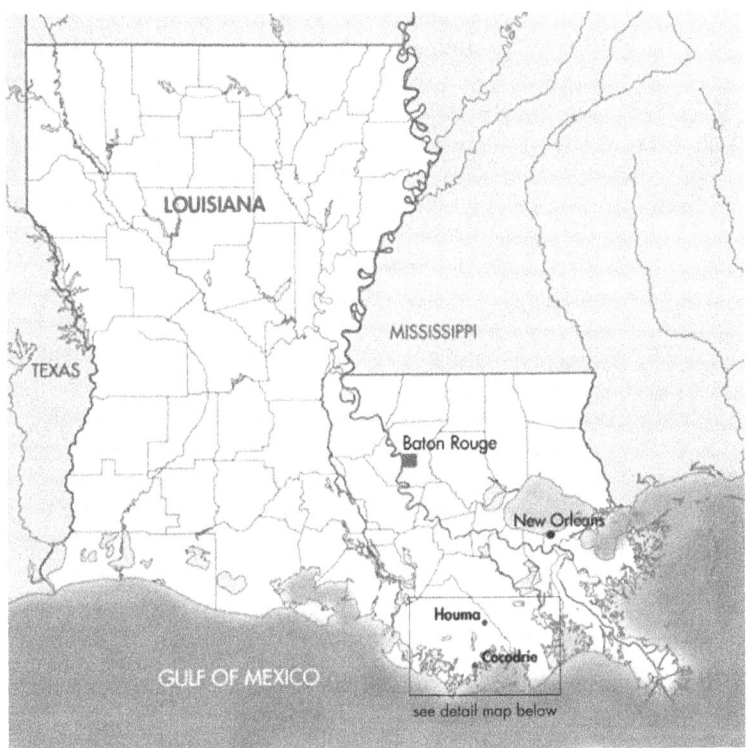

The British expelled the Acadians from Nova Scotia in 1755. Within a few years, many of the French-speaking exiles had worked their way down to France's colony on the Gulf Coast, where they settled in the coastal prairies southwest of New Orleans. The Cajuns lived off the land, farmed, and in time developed an economy by shipping seafood and fur into the city.

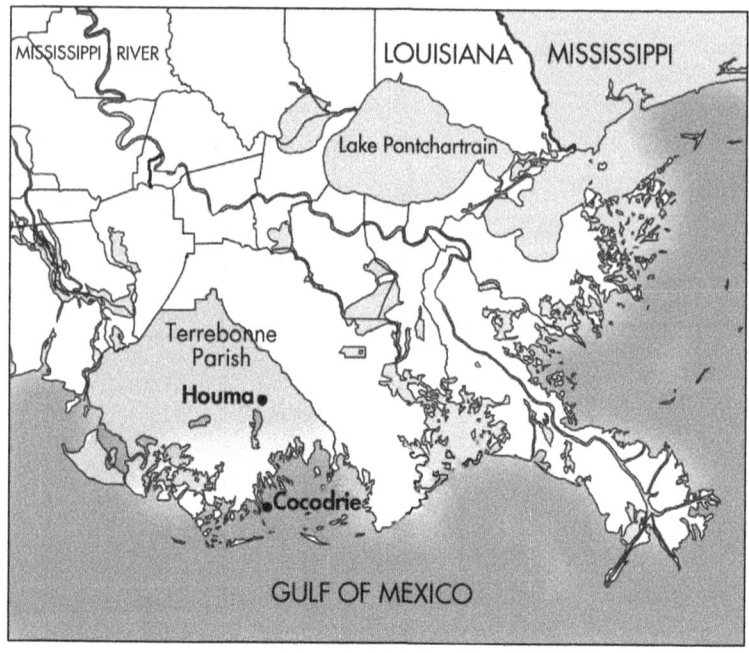

Cocodrie, on the coast of Terrebonne Parish.

Today, the stronghold of Louisiana's saltwater Cajuns lies within Terrebonne and neighboring Lafourche Parishes. Terrebonne is the larger, but both of the marshland parishes face the same fate: With their life-giving bayous dammed off from the Mississippi, their southern reaches are crumbling into the Gulf.

Kerry LeBauve, with his father Randolph, long represented Terrebonne Parish fishermen up in Baton Rouge. Kerry lives down in Cocodrie, which is Cajun for "alligator." The village lies at the end of the road, as far south as you can drive in Terrebonne Parish.

The first time I visited Kerry, it was just a few days after the net

KERRY LEBAUVE COCODRIE

Kerry LeBauve aboard his lugger.

ban had taken effect. His roofed dock along Bayou Petit Caillou, where he keeps his fishing boats and gear, was freshly littered with a collection of barnacle-encrusted stumps, roots and limbs.

Kerry, with his brother-in-law and sidekick Glenn Daigle, had been enjoying a fine winter, with plenty of fish and good prices. Then came February 28, 1997—the last day they could legally net good edible fish—and Kerry videotaped the big trout as they came aboard, "For my grandkids to see what I used to do."

With their incomes suddenly interrupted, the two fishermen threw out some crab pots. But crabs are scarce late in the winter, and things weren't looking too pretty.

It's one thing to work hard for a just reward, and it's quite another to bust your gut for nothing. But if Kerry was disgusted, he wasn't showing it. Instead, this lifelong fisherman was busy redirecting his energy toward new ventures that didn't have anything to do with food production. Which explained his collection of drying "whatnots."

Stands of great live oaks had once lined most of the major marshland bayous. But after flood-control projects put a stop to the flooding that, each year, had deposited a sustaining layer of silt upon them, the bayou-side ridges began to sink. The trees drowned and slowly fell, piece by piece, into the water. Oysters and barnacles grew on the smooth wood, and the textured bottom attracted fish. You could gently idle a boat through these stumpy minefields, and lay your net around the fish, but the snags played hell with your webbing as you hauled it back.

With the fish soon to be off limits, however, Kerry and Glenn began to see the relics of the old trees in a new light: Each time the net hung up on a gnarled stump, instead of cursing it, they gently rolled it into the boat.

KERRY LEBAUVE COCODRIE

Rich in tannic acid, the oak pieces were dark as strong South Louisiana coffee, and after a scouring with a pressure washer, the bleached barnacles that grew on them looked like patches of snow.

"They're not gonna run me off this bayou," said Kerry, who planned to continue making a living off the water by selling his "whatnots" to taxidermists and craft shops.

With the help of a filmmaker friend, he was also putting together the first of a series of videos that he hoped to market to sport fishermen. His "Videomaps" would offer the sports guided tours of the area, by Kerry himself.

The irony of filling his old fishing holes with sport fishermen wasn't lost on Kerry who, grinning mischievously, explained, "Yeah, I wanna fill 'em up. See, there's only one thing a sport hates worse than seein' a net fisherman in his fishin' hole. And that's seein' another sport!"

That was about the bitterest thing I'd ever heard from Kerry, who quickly added, "Don't worry, I'll be talkin' to 'em too, you know, about the cause." The "cause" was, of course, net fishing.

Kerry still hadn't lost his faith in the cause when I visited him for this interview, 3½ years later. But he'd given up on the whatnots and videos and, except for fishing some minnows for the sports, he'd fallen back on what he did best—feeding the people.

It was two days before the August, 2000, white shrimp season was to open. Kerry's boat was ready, and he had some time to kill. His wife Donna, who kept track of the family's fishing business on computerized spread sheets, was there too. But brother-in-law Glenn was off on a shakedown cruise, testing the new 3208 Caterpillar that he'd just put into his own Lafitte skiff.

Here, Kerry describes growing up in a fishing family, in a marshland community that's growing more threatened each day by a double whammy of environmental degradation and inequitable fishery management:

I'm 42, and I'm a fifth generation commercial fisherman. I'm the fifth generation, and the last of it.

I got three brothers and three sisters. Now, one of my sisters is married to a commercial fisherman. That's brother-in-law, Glenn. And one of my brothers, little Terry, is a commercial fisherman. That's it, just us three. My older brother, he's a shipyard man. He took welding at the vo-tech, and when he got out of school, he went straight to the shipyards. And he's been there ever since.

But even while he was in the shipyard, he always played around with the shrimping. He had him a little old lugger, and then he got a Lafitte skiff. And he bought him a little gill-net boat, to do a little bit of fishing. But the money wasn't steady with it, so he kind of swung back to the shipyard. He finally realized that if he wanted to go play around with the shrimp or fish, all he had to do was come on the boat and ride around with us a little bit, and get his fill. But he would've made a good fisherman, a real good fisherman.

I got three kids. I got a girl 21, she don't live at the house no more. She got married in April. My older boy, he's 17, graduates this year. He wants to be a carpenter, so I'm encouraging him on that. And my youngest, he's 14. He's not too sure what he wants to do, but fishing would be down his alley if I would push him. I wouldn't have to push him, no, all I would have to do is just take him and let him come do a little more. He's just natural with it, he likes it.

You gotta like it. If you like it, you can make it. If you don't like it, you ain't gonna make it.

KERRY LEBAUVE COCODRIE

He'd be a good fisherman but I ain't teaching him how to do it. He's already starting to get the knack of it a little bit so I don't bring him with me. He might get to like it too much.

That lady, Miss Vinton, at Southern Mutual Help, she tells me, "Why don't you teach your kids, somebody's got to teach them, to keep the future going. People are gonna need to fish to catch food for people."

But they don't want us around. So I'm not gonna teach 'em to do something that there's no future to it, you know?

He wanted to stay home on Monday to come shrimping with me, and I could use him. But they just started Friday, and these days right now, you miss four days of school and don't have a doctor's excuse, they want to suspend you. There's no, "Well, I need him at home." That excuse won't cut it.

But that's how I was raised. I stayed home to go shrimping and my daddy wrote a note, "Look, Kerry was needed at home." If they had any problem, they'd call up and talk to my daddy. And he'd tell 'em, "I need him to come deckhanding with me."

The first week of the season, you need help, to catch the first charge of shrimp, to hit that first week hard. If they didn't start the school year 'til a week after the shrimp season started, all these kids up and down the bayou could go shrimping with their daddies, make their opening day with them, to make a few dollars and help their daddies help their families out. Because that's what it's all about, it's family operations, you know?

When I was going to school, I used to always miss the first two, three days, for the August season. And the last two weeks of school there, the May season would start up. I would go to school and I would get a release out of school two weeks before everybody else.

LET THE GOOD TIMES ROLL

To make the shrimp season. Oh yeah, I never made the last two weeks of school.

When I was in the fourth grade, one of my teachers said—he said it in front of the class and I was kind of embarrassed about it, but it kind of made me feel good too—I had missed five days of school because we had went out to the beach and bad weather caught us and we couldn't come in. I got in Friday, we had a test, I took the test and I made an A—the only one in the class that made an A. The rest of 'em's all in school all week and didn't make no A, and the teacher says, "I don't know about y'all but Kerry must be studying his books behind the wheel of that shrimp boat."

We'd bring the teachers some fish, we buttered 'em up, you know, bring 'em a few shrimp. And we talked with the principal, and as long as I kept my grades up I was able to have that. I've seen myself miss 25, 30 days of school. But I made A's and B's. School was easy for me, I didn't have no problems. I mean, it's simple—you go to school, you pay attention in class and you didn't have to study. You just listened to what the program was about.

My older brother, he wasn't fortunate enough to get as many days off as me because he didn't make as good grades. So he had to stay going to school, and I got to go fishing and shrimping.

When I was in South Terrebonne High, we had the vo-tech. I took Nautical Science at the vo-tech. It was more for the oil industry, so we went there, me and my daddy, we went to the school board and we got together with them, to get us a commercial class set up over here. And the school board sponsored him to go to South Carolina and study some of the programs they had.

When he came back, he said, "Man it's unreal. The whole

community revolves around the commercial industry." He said they got the classrooms right there on the waterfront and the people go out on the water and fish while they're in school, to teach them how. The whole course revolved around the fishing industry, from the mechanics of the ice makers and all that, to the boatbuilding, the netmaking, to the processing of it.

And that's what we was trying to get started over here—a commercial fishing vo-tech school, to teach students how to be fishermen.

You know, we was looking for the better of it. We studied Florida, we watched Florida. That's what our dreams was, that Louisiana could be like Florida one day. Well, how Florida used to be. You know, anything in the water you could fish and sell and make a market for it to make a living out of. Because this is a commercial fishing area, especially on these bayous right here.

But the school board didn't want to back us for the funding, and this and that, and the insurance part of it, to get the insurance to bring the people on the boats to fish and shrimp. A lot of red tape, in other words. And they didn't wanna favor for it, they was against us so... .

My daddy was a shrimper, well, my daddy was just a *fisherman*, period. When he was a little boy, he lived right up there where the Intracoastal passes through Houma. He started fishing when he was a little boy, trotlining for catfish in the Intracoastal. And he would sneak off and come down the bayou and go shrimping and fishing redfish and stuff. He had one of his great-grandmas that lived in a palmetto hut just about half a mile up the road over here. He used to come down and hide at her house and go fishing with them and stay in the palmetto hut, with lanterns.

LET THE GOOD TIMES ROLL

They would use big cotton string and bend a nail for a hook. They'd go on the other side the bayou and catch all the fish they'd want. That's when wintertime used to get cold. They'd catch their fish and just hang them up outside their house and they'd stay frozen for two or three days.

And that's how he got fishing into his blood.

He was one of the last people that fished with seines, the drag-down seines and stuff like that, on the beaches. He lived in Florida for years, and ran charter boats out of a couple of different spots, down by the Keys.

He was one of the first offshore charter fishermen that worked out of this area right here, out of this Sportsman's Paradise marina. They had an old wooden crewboat named the Patty Ann. And he used to take people offshore fishing with that. Oh, that's in the early and late '60s, for any kind of fish. At the time, snappers was popular. Groupers, amberjacks, all that stuff. He trolled for kings, he did just about all you could do there, with the fishing.

In the summertime, when I was growing up, my daddy had a little lugger like this one right here. We'd shrimp. And during the split between the May and the August seasons, we used to go out with the line, with the outboard, every day. We used to get sick of fishing with the rod and reel. We'd go out there and catch a couple hundred pounds of trout every day with the line, and go back and sell it.

We used to fish in the winter months with trammel nets, but in the summer months we couldn't use the nets, because we didn't have an efficient net to use—too many catfish. So we used to go out with the line and just sport 'em. For the fun of it, plus make a few dollars until the season came around.

KERRY LEBAUVE COCODRIE

Except during the brief mullet season, LeBauve's net boats stay mothballed ashore. Note outboard engine situated in well toward bow of each skiff.

Then, in the early '70s, the Florida fishermen came to Leeville. When they came to Leeville, they introduced the monofilament net. And we got hooked up with them, and we got us a monofilament net, and that gave us a tool to fish with during the summer months. When that next shrimp season came up, me and my older brother told my daddy, "Oh man, let's not go shrimping." Shrimping sucked, you know. "Let's not go shrimping, let's stay gillnetting."

So we went and got us a gill-net boat, with a little motor in the front. We was using putt-putts then, little Briggs and Strattons, and that didn't work too good, it wasn't efficient. You had to take your lugger out to the beach, tow the little boat, and putt-putt around in that for two or three days to make you a trip.

LET THE GOOD TIMES ROLL

With an outboard, if the weather was nice, you'd just get in your boat, go out, fish that day and come back that day. And if the weather was bad, you weren't stuck out there for two or three days. And then you was able to cover more territory, do a lot more fish hunting.

And that was the last year we shrimped. My daddy let somebody run the shrimp boat for a season or two, and then we took the shrimp nets off and just used it basically as a standby boat when we made trips. And we eventually got rid of it because we didn't need it, after we started going with bigger engines.

We started off gillnetting with little 25-horse outboards, 18s, old antique motors, stuff you had to put wrenches on 'em all the time. And boats no bigger than 15-, 16-foot, with 12-inch sides on 'em. We'd go across Lake Pelto, head to the islands, fish in the surf, five-foot swells and all, with those little motors, and it was hard. Then later on, we got us a bigger boat, 21-foot, with a 115, where you could get out there and put in a thousand pounds of fish and still get in kind of fast. We got more efficient.

At one time we had up to seven gill-net boats that we owned and operated for ourselves. I was running the fleet with a crew of about 14 men. My daddy used to fish with us, but he got to where he was staying home and just marketing the fish. And I was running the crew.

I would go and pick the deckhands up, come back, and we'd get in the boats. We'd head out at 2 or 3 in the morning, then we'd come back in, most of the time, for 3 or 4 in the afternoon. All the fleet would come in, we'd have like 10 or 15 thousand pounds of trout, and we'd all gut 'em and ice 'em down. We had like our own fish house, because nobody was buying fish down here then.

A lot of times, I would take one boat and go out, load it down, come

KERRY LEBAUVE COCODRIE

back and jump on another boat and go back out with another crew. I'd wear a crew out. I had to shift-work deckhands, I'd bring a deckhand out there with me 'til he couldn't go no more, I'd bring him in, drop him off, get another one, turn around and go back out.

It would take an hour to get out there and that's when I would sleep. I'd let them take the wheel, and I'd take an hour's nap. Once we got out there, we'd start hunting the fish, and when it was time to come home, they'd drive and I'd get me another hour or two nap. I'm one, I can charge my battery up quick. After two hours sleep, I'm good for 10, 12 more, because that's how I was raised, that's how we fished when the fish was coming in hard.

A lot of times, we'd camp out on the beach. That's what I used to love about the fishing, going out by the beach. We'd go out there with our sleeping bags and make us a bonfire. When the tide wasn't right, you took you a nap, when the tide came around, you got in your boat and you went fishing. You might end up on another beach for the next night, you was all the time running around.

Out there, night and day is so different, it's a whole different world on the water, from night and day. You go out there on a day like today, hot and calm like it is, it'd be clear and you won't see hardly nothin'. No fish, no activity, everything is just, it's pretty, but there's nothing happening. But when the sun sets, it's like a whole different world. The water'll light up with the phosphorus, you see all the different kinds of fish swimming in it and kicking around.

It's funny how the fish come in. Where do they come from? All day, you try the deep water, you try here, you try there, you try everywheres to find out where the damn fish are at. Nothin'! But, boy, when dark would come, when a certain tide would change, you can't miss

LET THE GOOD TIMES ROLL

it, you throw your net *anywheres* and you catch fish. Where do they go, where do they come from?

It's something that as a fisherman you're always trying to figure, and that's what keeps you driving. "They gotta be somewheres!" Man, you try by the rigs, you're all the time trying something new.

But the shrimping, that's why I don't like shrimping—it's too routine. You know just about all you need to know about shrimping, all you gotta do is just go out there and grind.

If I'd have known that I'd have to be shrimping now, I never would've got into it. I'd have went to college to be a marine biologist or something. I'd go join Jacques Cousteau, something more adventurous.

I rate my fishing industry in categories, from top to bottom. The top is gillnettin' and the bottom is oystering. When you fish oysters, that's rock bottom! Gillnetting was top, and under gillnetting was shrimping, and then crabbing. I didn't do much crabbing but I had to get into that. Now we're into the crawfish. Crawfishing, I bump that up to right below the gillnetting now. Gillnetting, crawfishing and then the shrimping.

My great-great grandpa used to be a shrimper. In them days when you would shrimp you'd go for food vouchers. You didn't get paid. When you came in, you sold your shrimp at Chauvin's store, or St. Martin's, and they gave you a food ticket. And you was able to go get so much food out of the store, for your share of the shrimp.

There was very little cash back then, and not that many people had their own boats. You worked for a company and they each had their own fleet, like St. Martin's fleet, or Chauvin Brothers fleet. They'd each have eight or ten boats, and they had their own canning factories.

KERRY LEBAUVE COCODRIE

Kerry sorting shrimp catch.

That was up to, and during the war. Then, toward the end of the war, and after that, Texaco came down here, and it gave these people a chance to earn *money*. So a lot of these people raising families, they got off the boats and went into the oil industry. That's what Donna's family did. They're from up in Chauvin and just about all her family's related to the oilfield, her daddy and his brothers and all. See, that's the Texaco generation.

When Texaco first came and started drilling down here, that's what took that generation. That's how they became the oilfield generation instead of the commercial fishing generation.

LET THE GOOD TIMES ROLL

The main bayou that runs through Houma, that was like the main street for the town, like Canal Street in New Orleans. Down here along the bayous, all you had was some shrimp-drying sheds. The heart of the activity was in Houma.

Everybody'd take this bayou right here to go to Houma, to go to the canning factories, ice houses and all that. That Waterlife Museum that they built in Houma? That's where that originated from. These luggers, the shrimp boats and fish boats and oyster boats used to work down here and chug-a-lug all the way up to Houma, with those old gasoline engines. To unload and get supplies.

The waterways was their main route, and the highways was just dirt roads.

My grandpa, when he was a little boy, used to come down this road right here when it was a dirt road. He'd take it all the way out to what they called Point Masse, in a horse and buggy. Then he'd take a pirogue and paddle to Last Island.

But Point Masse, where he used to park the horse and buggy, there's no more of that—it's in the middle of the lake right now, underwater. It's about two or three miles off the bank.

We got some of the worse erosion in this area right here than anywhere in the state. It's bad, you see it going every day. Saving the marsh? It's all good, but it's too slow. The process that you got to go through, the red tape before you start a project, it takes too long.

The funds are being used for studying it, instead of doing it. Millions and millions get appropriated and instead of putting it right where it needs to be at, it's going to all the people in the middle to make their money. The studiers.

They got a guy in Houma, man, he's set now. He's been studying

KERRY LEBAUVE COCODRIE

Before there were highways, Terrebonne Parish's seafood harvest was distributed to distant consumers from the inland port of Houma. The main artery through Houma was Bayou Terrebonne, shown here in 1934. Bayou Terrebonne is a branch of Bayou Lafourche, which in turn is a tributary of the Mississippi River. When Bayou Lafourche was dammed in the early 1900s, the natural hydrology of these waterways and their tributaries—like Cocodrie's Bayou Little Caillou—was drastically altered. In the 1950s, the U.S. Congress declared Bayou Terrebonne a non-navigable waterway when Houma's town leaders expressed a desire to fill it in to create additional parking.
(Terrebonne Parish Consolidated Government photo courtesy of Thomas Blum Cobb)

the marsh for years and years. He's *still* getting checks for studying it, and they still ain't started 90 percent of his projects. Everybody could see that. I mean it's getting to the point where you're tired of hearing about it because it's all talk and funding but no action.

See this ridge of trees right here, on the other side the bayou?

KERRY LEBAUVE COCODRIE

That's the only thing that separates us from Terrebonne Bay, and it's sinking and the trees are dying.

If you get right on the other side of the trees, it's all flat marsh, with one or two little canals, then you jump to the next bayou, Bayou Terrebonne. And that's so washed away right now, on a high tide you got to watch how you run down it, or you'll go off into the marsh, not realizing you're not in the bayou. That's how little of the marsh there is left.

If they wanted to save this bayou right here, they'd have to come back, build this all up into a good high ridge, and plant some trees on it, trees that grow fast. Willow trees, stuff that can grow quick, get big, and give a foundation to that levee. And then you'll have something, because when we get a storm, all our high water comes through this ridge. When Andrew came, right here where them trees are broken up? A tidal surge came through right up there, pushed all these camps up against the highway and it came up and gutted out a channel right through here.

And every storm we ever had, that's where we get it from. A southeast wind takes Terrebonne Bay and pushes it in right here. And you know how big Terrebonne Bay is?

[Opposite, above] Situated at the convergence of six bayous, Houma was founded as Terrebonne's parish seat in 1834. A parade celebrating the town's centennial featured this float which demonstrated the seafood industry's importance to the culture and economy of the region.

[Opposite, below] Another float in the 1934 parade marked the recent addition of the petroleum extraction industry to the region's economy. Inherently non-sustainable, the industry eventually eclipsed the area's traditional fishing industry.
(Terrebonne Parish Consolidated Government photos courtesy of Thomas Blum Cobb)

LET THE GOOD TIMES ROLL

As long as that ridge keeps going down, you're gonna have more and more damage, and less and less on this side of the bayou.

Well, they already said the federal government wants to stop spending money on these little areas. And it does make sense. I mean, if they ain't gonna save it, if they ain't gonna keep it up, it don't pay to keep putting money into developing it.

But the only problem is, the poor people don't have the money to stay down here, but the rich people do. So that's what's gonna happen. The average working-class people that's working down here, that's living down here, they're gonna have to move off because they won't be able to afford the insurance.

Us, we lose this, we lose everything. We got to start our business all over back from scratch. But, for the sports, it's a second home to them. They lose that, what do they lose? Nothin', because the insurance pays for them to build a bigger and better one.

Sports that are broke down? I still bring 'em in. Sometimes they wanna give me money but I tell 'em, "Hey, next time it might be me."

But one time, they called me up in the middle of the night, got me out of bed. The marine patrol's boat was broken, or they didn't wanna go, and they called me up and said there was a sport broke down in Bayou Cantrelle. There was four people in the boat, and it was pouring rain. So I said, "Okay, I'll go get 'em." I mean it was pitch black and pouring. And when I got in the little boat and left, it just started to rain harder. It was miserable.

So I find 'em in the night there, and tow 'em in to the launch. And when I was getting ready to leave, the guy slips me a bill all folded up. "Here's something for your trouble," he said. I didn't even look at it,

KERRY LEBAUVE COCODRIE

Tourists with bull reds at a Cocodrie sport-fishing marina.

until I got home. When I got inside, I took it out my pocket and looked at it. Five dollars!

It was so much trouble, man, it should've been worth a hundred bucks, at least 20 bucks would've been saying something. Or nuttin'. I'd a took nuttin', a big thank you. If he would've said, "I appreciate it," I'd've walked out feeling better. But when he gave me that five dollars, it was like he just slapped me in the face. "Something for your trouble!"

That's why, when people say, "Oh man, go in the charter business," I say, "Been there, done that, don't like that."

We already had a marina. Ninety percent of 'em are good people. Nine out of ten of 'em you're gonna bring on a charter trip, you don't find no better. They come down with the right attitude, they're friendly. But, boy, you get that 10 percent, they come down with that

Kerry LeBauve's lugger, rigged with skimmer nets which scoop up shrimp as they ride the falling tide.

attitude that they're better than you, that you're their nigger. And they treat you like that—you got to *serve* them, because they're paying for your services. And one like that'll ruin your whole week.

We used to take people charter fishing, a hundred dollars a trip. If you didn't catch fish, you didn't pay me. If you caught fish, you give me a hundred dollars. And we got paid every time.

But that was when I was into sporting a lot. I knew how to catch 'em with a hook and line. I'm out of hook-and-line practice now. It'd take me another season there to get back to learn how to catch 'em with a hook.

KERRY LEBAUVE COCODRIE

But I don't like hooks, they're sharp, they're pointy, they hurt when they stick you, they're dangerous, man. Give me a net instead, they're soft. A catfish might get me taking him out of the net but at least I don't get that hook stuck in my hand. Like in that show, "The Perfect Storm," that old boy went overboard with that longline? And they had to be cutting that big hook out of his hand, ahhh!

I didn't even get the rod-and-reel trout license. It's not profitable, because the time of the season you got to fish 'em is toward the winter months. We knew that when they put that law into effect. We asked them to give it to us during the summertime. If they'd've gave us a rod-and-reel license during the summertime, I'd've bought me one. I'd've went trout fishing in the summertime with a rod and reel and I'd've *made* money. But in the wintertime, it's a joke.

I heard some people were trying to change it, but for me, there's no reason to go up to Baton Rouge anymore. If the issue comes up, we don't go anyhow because it's pointless. They don't listen to you. It ain't good but what you gonna do? That's why we're just diversifying and going with the flow.

We bought the lugger, a 42-footer, the year before they passed the law, because we seen it coming. And I'd been watching to get a boat because, people don't realize it, but before the net ban was in effect, we had bad years on the fish, freezes, whatever. I *had* to go shrimping a lot of times. You'd get a season where they didn't have much trout, or the fishing was slow, and it was good for shrimping, well I would go on a boat and deckhand for one of my friends for a while.

And when we seen the net ban was coming around and we was fixing to get put out of business, well, it was time, we better get us a shrimp boat *now*, for when it happens, we'll be ready, we won't be out of it, you know?

Then I turned around and bought some crab traps. I went crab fishing after, because the shrimp season only lasts so long. So you needed something to fall back on for the winter months.

Crabbing ain't bad, it's alright. It's just like anything else, it's not the best, it's not the funnest thing to do.

Minnow fishing, that's pretty fun. I was doing that right now. I just shut it down because it's time to go shrimping. We get 10 cents apiece. We got a couple of bait shops up and down the road over here, and I got one I sell mainly to.

You can catch a lot at the right time. Everything has its cycle. That's one reason why I shut it down—when it gets toward this time of year, the minnows don't bite as good, they start layin' their eggs, they don't feed as heavy and it's so hot. So you have to go out there with twice the traps, and fish twice as hard to catch a few.

Last year was our first year with the minnows, but it looks like anytime we start with something it goes down. I made 25 traps for last year and I didn't even put 'em in the water. They had so many minnows, anybody could catch 'em anywheres. They didn't want to buy 'em, so I didn't even wet the traps. This year here it was a different story. Everybody wanted 'em, so that's when I went minnowing.

We got into crawfishing because we needed to do something in the wintertime, when we used to catch reds and stuff. We heard about it for years, the crawfishing, and talked to crawfish fishermen. They said, "Man, you can do real good. You can make you 40, 50 thousand in a season, 30 or 40 thousand dollars in three or four months. To us that sounds great. I mean, even if you don't clear that, if you only clear 10 or 15 thousand, for two or three months, that's still good money.

A buddy of Glenn's had a friend that was a crawfish fisherman in Pierre Part, in the Atchafalaya Basin. And when he started getting ready to go, Glenn and his buddy started talking about wanting to go too. We batted the idea around awhile, and we figured we'd go ahead and give it a try. So we went and bought us some wire.

We didn't have the boat to do it. We just had a regular old square-fronted flat boat, but we went over there and we gave it a good try. We did alright, for beginners. We talked to them fishermen and they told us it was the worst season in 25 years. And we were happy. We got up to 10, 12, 15 sacks a day. And these old boys, they was coming in with their 10, 12 sacks too, so we figured we was keeping up with the average fisherman. We was doing something a little right, you know?

But we lucked out, we happened to go over there and throw over right into the crawfish.

We didn't know what we was doing, we threw some traps here, we threw some there. We didn't even know how to get around the swamps. I mean we was lost, as new to it as you could get.

But right away, we had crawfish in the traps. So we started fishing 'em, we started throwing our traps in circles. We're thinking it's like crabs, you know? "They're biting right here, so you pile some traps up right here."

That's not how you're supposed to work 'em, but we started catching a few anyway. Then, next thing you know, man, all the other fishermen started coming in where we was at. That's when we realized we was in the crawfish and didn't know no better.

They started putting traps in there, and crossing it with long lines. But they know the swamp, they know where the ridges go at, the high and low levels, and all that. And they would come and shoot lines from one end of the swamp to the other end with 50, 60 traps. And then come back with another 50, 60—long stretches, where you could make time, because you got to travel through that swamp. It's a slow process, hitting trees, getting caught on logs, working your way through there.

But, like I say, we'd go in there, man, and put traps in circles, and then some thieves came through and made a raid one night and we lost about 20 of 'em. You don't want to stack your traps up like that because when they come there and hit you, they're gonna clean you out for everything.

We had just learned that we gotta make long lines when the water got bad and the crawfish was dying in the traps. We didn't have big enough traps. When you first start fishing crawfish, and the water's good, you can leave your traps under the water. But once the water starts changing, when you start getting bad water, with no oxygen, you got to have your trap out the water, so the crawfish can climb up and get air.

At the time, all we had was some four-foot traps, and the last time we went run 'em we had like seven or eight sacks of crawfish and just about all of 'em was dead.

What the real crawfish fishermen was doing, they was pulling their traps out and adding extensions. Making six-foot, eight-foot traps, so they could lean 'em against a tree and let the crawfish climb up 'em and get air. If we wanted to stay with the crawfishing we would've had to pull all our traps, add extensions to 'em and turn around and put 'em all back out. And the shrimp season was coming around, so it was like, "Well, it's time to shut this down 'til next year."

The first year we started, I knew I was going back. And I needed a boat to crab with, because I had this old fiberglass thing and an old aluminum hull that was all bent up. I had bought a new motor so I figured, "Well, if I'm gonna be able to pay for my motor, I'm gonna get something I could use that's versatile, something I could crawfish with, crab, minnow fish, all the way around."

So I got this old boy in Pierre Part to build me one of those Basin boats, with a pointy, turned-up bow. By the time he had it made, it was time to go pick the traps up. So I got the boat from him, we bolted the motor on, launched it, and went pick the traps up. Then last year we went four or five times, and that was it, because last year was a bust.

We made our mullet season last fall, and then I bought a thousand dollars worth of crawfish wire. Now I was ready, I was gonna go crawfishing with about 800, a thousand traps. Go do it *for real,* you know?

We was making the traps, and getting all excited about the crawfishing there. We had our crab traps in the water, we was crabbing, waiting to get the itch to go crawfishing. We were talking to a few fishermen over there, trying to get the report, and they'd tell us, "Oh, there's no water in the river yet, not right, not quite right."

It got around to Mardi Gras season and that's usually when some of these fellows kick off. So we said, "Let's go give it a try anyway. At least get our traps in the water, start getting set up."

It all takes time, it's like a two-hour drive to Pierre Part. And you can only put out, in a day, about 50 traps. You can only put so many in the boat and by the time you take and you stretch 'em out, you find a tree to put 'em on, clean the bottom, flag 'em off and you jump to the next tree, and the next tree, it's a slow process.

The first time we brought like a hundred traps, and we put out 60.

LET THE GOOD TIMES ROLL

And we went three or four days later and brought us another 100, and put out about 25, 30. We checked the 60 traps we had and I don't believe we had ten crawfish. "Ahh, this is not looking very good here."

So we only threw out another 30, 40 traps. I said "Man, let's hold off. We ain't gonna put out too much of this in the water, because if this don't get too much better, we got to come get these things out of here!"

We had about a hundred traps out there and we went a few days later and checked 'em all again, and baited 'em, and same thing—we didn't have more than about one, two crawfish to the trap, if any.

A lot of places we was working last year, you couldn't even get in because they didn't have no water. And that's when we seen, "Hmmm, this ain't too pretty." That was this spring, and the drought got us, the worst drought in a long time.

Then we talked to a fellow that's got a shrimp boat down here, that lives in Thibodaux. He was crawfishing, and he was making money fishing in some ponds in Chackbay. So we got the name of some people there that had some land, and we went fishing in the Chackbay swamps.

We got a lease that they let us fish, a mile long. And that was all walking, there was no boats for that. You'd take a pirogue and you'd pushpole. And you'd have to get in the water and walk through the cypress knees, waist deep in water, tripping over logs, dragging a pirogue behind you. We was working our butts off trying to make this thing work, you know?

We started off good over there, we got up to 12, 15 sacks at our peak. Then some bad water came in, and we got dead crawfish again. Then, after that, it just went down, the crawfish buried or something, the bad water did something with 'em. So we got out of there.

KERRY LEBAUVE COCODRIE

Taking the nets? People have no idea what happened. They still got their heads spinning, man, with the reality of it.

Every week they had a new porpoise in the net, so claimed. But it was the same picture, just at a different angle. They must've had about 20 shots of it, and they just showed one at a time.

Edwards, before he left office, he should've took and threw that all out in the garbage. And let them worry about it in the next session after. But he didn't back us up.

Edwards had his ass in a crack with the casino issues and all this other stuff. The big shots had him under the gun. He couldn't, well, he didn't help us out. And commercial fishermen was one of his strongest constituencies. Southeast Lou'siana is what carried Edwards.

Now Edwards was a crook, but let me tell you about Edwards. Edwards shared for everybody. He might've been a crooked politician but he took care of *everybody* at the same time. These other politicians are crooked and they ain't taking care of nobody but them.

That's why politics in Edwards' days wasn't bad. Everybody got something out of the pie. He kept the working man working. He wasn't looking to attack the small people because he knew where he came from. But these other cats, like Foster, he knows where he comes from, with a golden spoon, so he ain't worried about the little people.

We're looking at a *total* net ban, I'm telling you. We're gonna have the shrimpers coming out of the water. Sooner or later they're gonna put this to a state vote. They're gonna take and put a constitutional amendment and we're gonna be completely out of the water. We're gonna be like Florida.

Foster, a couple of years ago, he had a meeting in Thibodaux's

LET THE GOOD TIMES ROLL

Chamber of Commerce, and that specific thing was brought up. That was when he was trying to push his bill for that amendment thing, so anybody could put anything on the ballot. And he slipped up and made one little point about it, "Yeah, like the net ban in Florida there, with the amendment, we could do that," he said.

Wow, I couldn't believe he said it in public like that, on TV, but nobody else realized what he said. That was after we was already out of business, so I'm reading between his words as he's pushing this issue. He said "That's how they did the *net bans* in Florida." "Net bans," not "gill-net" bans, "net." So he's referring to the shrimpers and everything else. But that was the only thing he let slip out about it, in public.

Ah yes, if Foster has his way, he's gonna have it passed before he gets out of office and then the next "yes man" that comes in is gonna push the initiative to do it. Foster's already got everybody in the state under control, whoever's gonna win the election is gonna be a Foster man. Well, a GCCA man, because Foster ain't the big kingpin neither, he's just one of their little pushbuttons too.

That ain't no nice people there, that's just selfish greedy people that just don't give a damn about the little man at all. They'll squoosh 'em out before they'll put up with 'em.

The whatnots and the videomaps were good ideas but they just didn't pan out. The whatnots never kicked off too good, we had to shut down the business on that. It was called Boondock Art, but it never did go too far. They sold one or two, and that was it. We still got our whatnots, up in an attic.

And our Hang Our Way tape, the videomap, it didn't sell too good neither. Modern technology took over too fast for us, the CD ROMs.

KERRY LEBAUVE COCODRIE

It took us a year and a half to piece it together, and by the time we got that done, you could get CD ROMs that took you right there. They're not the same, they didn't have our knowledge to it, but... .

See, we're trying to find some financial stability, that way we don't have to play with the goddarned shrimp and we could just go around working to try to organize to save what's left of the industry. And till we get financial stability, we can't do it. That's the bottom line.

You'll starve to death if you try to go to Baton Rouge, go to these meetings, try to organize stuff, *and* try to make a living. So you got to do either one or the other: Make a living or fight for the cause. And fighting for the cause has got to take second place, because you got to provide for your family first. So that's why we was trying to get one of these things to kick off to where we could make money with it. That way we would be sufficient enough to keep the bills going, and then we could turn around and have meetings.

Because I still believe that you could do it, I believe we could get our fish back. Oh yeah, if you had the time there, in maybe five or ten years, you could change it. But it's a 24-hour job.

When we started with this thing a few years ago, if I wouldn't have had to go shrimping and working and all, it would already be pretty much on a good stage to keep it up.

When I first started with this issue, I grabbed all the oldest and biggest shrimp dealers on this bayou. I went to all of 'em, personally, and I said, "Look, I want to stay on this bayou making a living. Y'all are the old timers, you got the power, and I need y'all's help."

"What can we do?" they said. "We're about ready to retire. It's up to y'all, the younger generation to take over."

They asked me, "Well Kerry, what you want? Do you want to start an organization? We'll help you, we'll back you up with the paperwork and all."

I said "No, I don't have enough knowledge to operate an organization. That's not what I'm looking for. But we are lookin' for *somebody* that's got the time and experience that can operate and pull the fishermen together. And y'all are the people that got the knowledge and experience of it."

"Well, we can't help you much there." And I said, "Well, who can?"

"Go check with the Ward 7, the KC Hall, the Knights of Columbus, all these organizations on the bayou."

So I went and talked with them. And all of them, a hundred percent behind us. And the VFW represents maybe half a million people across the country. You talk about political pull, if you could get these organizations to back you up and support what you want to do, you're talking about being able to pull in some numbers.

You know how GCCA says they represent 500,000 sports? You could be saying your own numbers. They got VFW's in every town, they got the Ward 7 Club, then we got the Knights of Columbus, the Lion's Club, Shriner's Club, Woodmen of the World, this is all clubs that is already established. They give money to charity, they help out the community with the churches, people get in a bind, financially, for medicine, the elderly, they're into all that.

That's the kind of support you need, people that vote, people that understand the small people. People that eat seafood, appreciate good seafood and can't go out and catch it themselves

You got to hit all their regular meetings and bring up agendas and ideas, and they'll jump on the bandwagon and help you. They'll love to. The Ward 7 Club was where we used to have our meetings. Any time we wanted that building all I had to do was call 'em up to make sure they didn't have anything scheduled and we could have the hall for free.

That's what I was going for when I was trying to do it. But I was fishing too. I can get plenty of people to follow me, yeah. I can talk with 'em, and they want to get involved. But I'm just a little peon on the bayou over here.

During the fight, we'd go up to New Orleans or Baton Rouge, and these big fish dealers would be walking around with their suits on, talking about needing *thousands* of dollars for this and that.

And here I am over here, talking about getting a couple of cans of paint donated and a few sheets of plywood to start making some signs!

They wanted thousands of dollars for this and that. "That's what the Seafood Management Council's for, man, to get everybody together." Well, I figured, they got an idea, if somebody wanted to take charge, fine, I'm just gonna follow with them. And what happened? The same old thing, mud in your face.

They wanted me to start an organization down here and I know right off the bat starting an organization is more than just getting a piece of paper and saying, "I'm the president." I've been down that road many a time. And I wasn't going through that.

But I was supporting the industry, fighting for it, by just getting involved with the people, you know. Trying to get people to wake up, and aware of what's going on, and the ideas of that, how to get it rolling.

Down here, when I went to these meetings, a lot of people'd ask me, "Whatcha want? You want money, you need money?" People kept wanting to give me money, write me checks.

I'd say, "Oh no, I don't want your money, because if I take your money, you expect something from me. All I want is, Be there, get involved."

'Cause it's easy for somebody to give you a hundred dollars and forget about it, not to get involved, but say, "What you did with my hundred dollars?" Instead, I'd say, "Hey, come put your goddarned butt right here and get involved with it. And put in a couple of hours of work and then you can say what you did with your time, instead of what I did with your money."

Now, I got people donated me money for these signs that we had, "A Louisiana Tradition." I found out what they were gonna cost, and I had a hundred and some dollars I put in. Miss Helen gave me a hundred and some dollars from Southern Mutual. So we was gonna go and buy like $300 worth of signs. Then this crab buyer, I had him at one of these meetings and he gave me $700 and said, "Here. Go buy you some signs."

And when they gave me the check, it went straight to Houma Map and Production. I showed 'em what was what. People wanted to give me all kinds of money but I didn't take a nickel from nobody so I didn't have to be accountable to nobody for nuttin'. And boy, right after that, after I get these people to come to these meetings, everybody else that's got organizations, they come in and, "Oh man, we need money, we need money, we need money." And here I am talking a just-get-involved-movement, more than just "put in your bucks," you know?

And when it came to going to meetings off the bayou, well, this organization on this bayou didn't want to be with that organization on that bayou, this organization didn't want to be with this organization, and this and that.

And it all went to where it's at right now. Well, fine, I did my best, and I can still make a living.

KERRY LEBAUVE COCODRIE

But you got to have your feet in every part of the business, and whatever's left of it, by the time you retire, thank God you got it, because it would be hard to find a job before you retire. What could I do? I'm stuck, I ain't gonna be able to sell my shrimp boat, ain't nobody gonna want one.

I guess I'm just gonna have to sit down on the side of the bayou there and just wave at 'em as they go by. Do like the old man on the side of the street there, the bum, sit there with my cup rattlin'.

LOUISIANA FISHER FOLK

ROBERT FRITCHEY
GOLDEN MEADOW & LEEVILLE

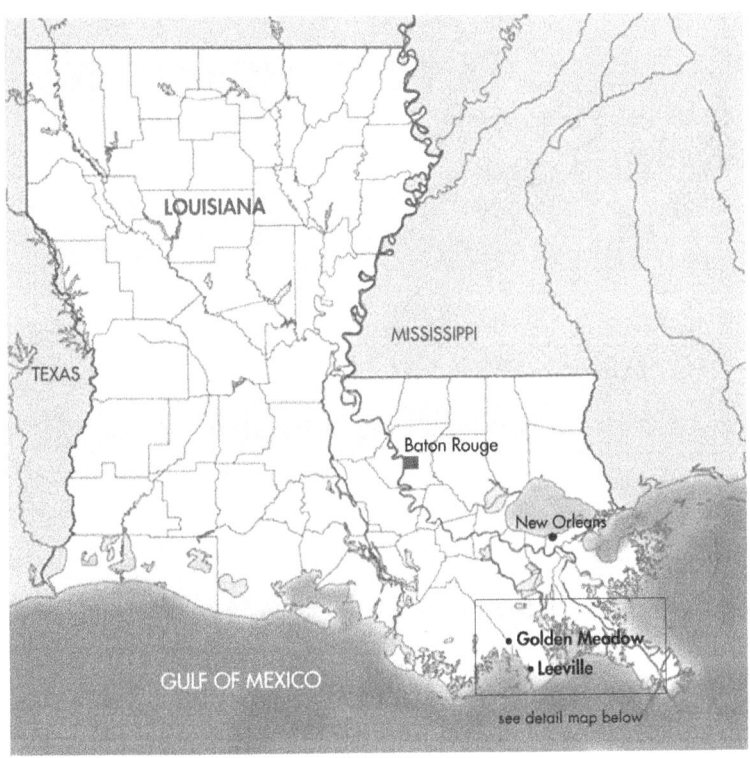

As its French name suggests, Bayou Lafourche is a fork of the lower Mississippi River. From its origin at Donaldsonville the tributary flows south for ninety miles through sugar cane fields, then salt marsh, to the Gulf of Mexico. Along the way it swells from a muddy backyard ditch to a tidal channel navigated by offshore vessels.

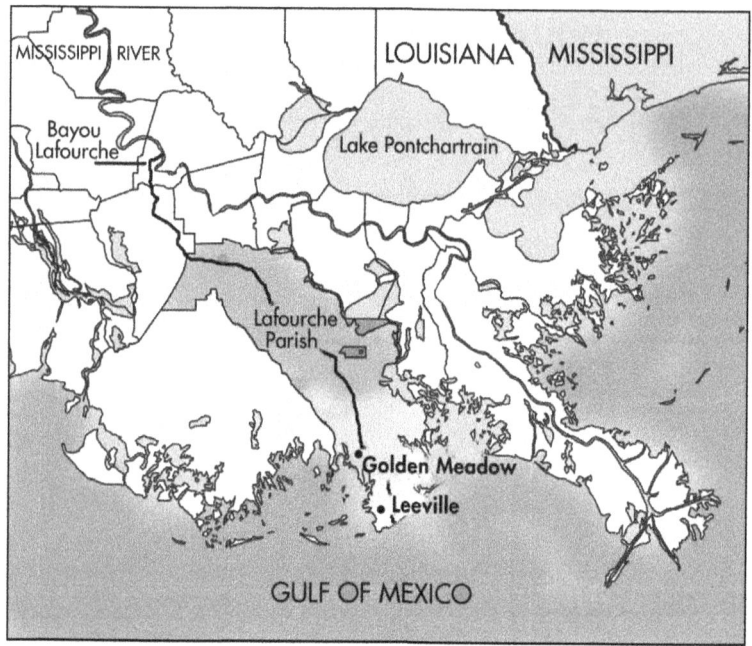

Golden Meadow and Leeville, in southern Lafourche Parish.

This is Cajun country and an uninterrupted chain of old settlements lines the bayou, until its narrow floodplain sinks into soggy marshland. The last link in the chain is Golden Meadow. Lying in the heart of Louisiana's coastal wetlands, the three-mile-long town is less than twenty miles from the beach.

Named for the rampant buttercups, ragwort and goldenrod that seasonally carpeted its pastures, Golden Meadow initially supported some farming. Now, with solid ground at a premium, the farmland's mostly been displaced by homes and businesses. Fishing continues, however, and has a long tradition.

ROBERT FRITCHEY GOLDEN MEADOW & LEEVILLE

Robert Fritchey *(Brian Gauvin photograph)*

Spotted seatrout on a drying platform in the 1920s. After anthropologists found bones of this southern species on the West Coast, they traced their origin to this facility, which was situated on an island lying inshore of Grand Isle—the cured fish had been shipped west to feed Chinese railroad workers. *(Ellender Collection photo courtesy of Thomas Blum Cobb)*

With gear they'd brought from Nova Scotia in the late 1700s, local Cajuns seined shrimp and fish which they consumed locally or boated to the city in live wells or layered in salt. Markets beyond New Orleans grew slowly until after the Civil War, when the region's abundant shrimp attracted the attention of Lee Yeun, a shrimp processor from Canton, China. Yeun introduced the shrimp drying process in 1873 and, by 1892, Chinese corporations had built six drying platforms in nearby Barataria Bay, manning them with Filipino laborers who lived in isolated villages they built on stilts over the bay. Cajun fishermen, traveling in lug-sailed *"canots"* and pulling their seines by hand, supplied the workers with shrimp, which they dried in the sun and exported to China, the Philippines and Hawaii.

ROBERT FRITCHEY GOLDEN MEADOW & LEEVILLE

Thanks in part to this early export market, shrimp landings, statewide, jumped from 534,000 pounds in 1880, to 7.2 million pounds in 1890.

Golden Meadow boomed in the early twentieth century after Biloxi seafood processor T.K. Devitt ventured down the bayou, scouting for a place with "plenty of shrimp, inside waterways, and easy access to the Gulf of Mexico." He built the community's first cannery in 1917, which encouraged entrepreneurs from Golden Meadow and New Orleans to follow with four more.

Between processing and canning shrimp and oysters, the factories employed hundreds, and workers flocked to the town. Paddle-wheelers hauled the canned seafood to New Orleans, where it was distributed locally or shipped out by rail to other states.

Seafood dealers also shipped out fresh product, iced in barrels. Later, advances in freezing and ice-making technology facilitated the wholesale distribution of frozen and freshly caught seafood to consumers, which ended the era of Golden Meadow's name-brand canneries. The town's next boom was in hotels and other businesses that catered to the *Texians*, who in the 1930s began to tap a non-renewable resource beneath the marsh.

The oil industry siphoned talent and capital from the developing seafood industry. Still, by 1949, according to an LSU economic report, Golden Meadow was homeport to a fleet of 300 fishing luggers which "each year brings in millions of dollars worth of shrimp and oysters, speckled trout, redfish, sheepshead and pompano."

Both the volume and value of the local harvest steadily increased through most of the twentieth century, though most of the action eventually shifted about nine miles further down the bayou, to Leeville.

Blessing of the Fleet in 1947: Fishing boats in Bayou Lafourche in front of Golden Meadow's Our Lady of Prompt Succor Catholic church. Note cypress cisterns behind homes in foreground, to collect drinking water. *(Courtesy of Daniel Lafont)*

Leeville's development was jumpstarted by an influx of evacuees who re-located inland after "The Great Gulf Storm of 1893" destroyed Cheniere Caminada, near Grand Isle. That Gulfside fishing community, which dated from the late 1700s, was just a day's sail south from the French Quarter in New Orleans, and had been the main source of the city's "fish, crabs, oysters, and shrimps," New Orleans journalist Catharine Cole wrote in 1892.

By 1905, 2,000 people lived in Leeville, where they grew rice and enough citrus that the town was, at that time, called Orange City.

ROBERT FRITCHEY GOLDEN MEADOW & LEEVILLE

The Petit Caporal—named after Napoleon Bonaparte—was built in 1854 of virgin cypress. Leon Theriot Sr. of Leeville purchased the lug-sailed vessel in 1899, and in 1902 converted it into the area's first motorized fishing vessel by installing a 3-horsepower tractor motor. Theriot fled the 1915 Leeville Hurricane in the "Little Corporal," and used it to relocate his family to Golden Meadow, where the vessel is currently displayed.

Fishermen hauled seafood from surrounding waters, and merchants maintained a brisk trade with the French Market via sailing vessels and, eventually, motorized luggers.

A pair of hurricanes early in the twentieth century forever altered the town's course. A September 1909 storm caused $5 million worth of damage and killed 350 people, statewide. Storm tides at Leeville destroyed the Catholic Church but casualties were light since most inhabitants had evacuated up the bayou in their boats. The storm discouraged a few residents, but most returned to rebuild.

LET THE GOOD TIMES ROLL

The second hurricane, in September 1915, killed 275 people and caused $13 million in damage, much of it in New Orleans. A 14-foot storm surge hammered Leeville, killed dozens, leveled most of the town's buildings, destroyed the trees, and carried off the topsoil. After that, all but a few contrary fishing families abandoned the isolated village to settle in Golden Meadow and communities even further up the bayou.

Leeville remained a ghost town until the early 1930s, when oil was discovered there. By 1938, nearly 300 derricks towered over the surrounding marsh.

The same year, Eddie Martin built the first modern shrimp shed in the port, shortstopping fishermen on their way back to Golden Meadow. As additional dockside buyers followed suit, so did fuel docks and icehouses, and Leeville developed into one of the nation's major seafood centers.

Since a few buyers and processors continue to operate in Golden Meadow, the two towns are linked together by the National Marine Fisheries Service, in its annual ranking of the nation's top 60 fishing ports. Dollarwise, Golden Meadow-Leeville ranked sixteenth and seventeenth in the nation in 1994 and 1995, when fishermen landed, respectively, $30.1 million and $31.3 million worth of shrimp, crabs, oysters, inshore and offshore finfish. In 1997 and 1998, as the net ban began to take effect, the ports fell in ranking to eighteenth place ($24.8 million) and to twenty-third ($27.1 million), respectively.

To control flooding, Bayou Lafourche was dammed off at its mouth in 1904. Just enough river water is pumped across the dam to maintain a gentle current. As intended, the floods ended. So did the annual introduction of sediments that sustained the coastal marshes.

ROBERT FRITCHEY GOLDEN MEADOW & LEEVILLE

Shrimp skimmer boats leaving Leeville for a night's work. Though it's regularly inundated by hurricanes, Leeville is ranked among the top seafood ports in the nation, in terms of money paid to fishermen for their catch. Laws that curtailed the harvest of red drum and other inshore species of finfish markedly reduced that value. Note marshland on opposite side of bayou. *(Brian Gauvin)*

For storm protection Golden Meadow and its neighboring South Lafourche communities—Galliano, Cut Off and Larose—surrounded themselves with a high earthen levee. But outside that bowl, the protective wetlands are reverting to open water, and these towns are becoming increasingly vulnerable. On its own—outside the levee system and nearer to the Gulf—Leeville's fate is even more precarious.

There are more boats than people and those few hardy individuals who persevere do so in homes perched high above the marsh. Waterfront seafood buyers and marina operators don't have that luxury, and their operations are regularly inundated by the floodwaters of passing storms, and bludgeoned by their winds. Afterward, they wash out the mud, rebuild, and go back to work.

This marshland port endures in a hostile environment. Corrosively salty winds blow relentlessly; when they do lay down, clouds of gnats and mosquitoes descend. Winter can be bleak and bone-chillingly damp; in summer, even the water is hot. Yet after centuries of commercial fishing, the surrounding marsh, like a bank, holds a fortune's worth of fish, and that's what lured me to Leeville:

Unlike most of the folks profiled in this book, my commercial fishing heritage doesn't go back generations or start in Louisiana. Dad was a doctor, a dermatologist with his own practice in Pennsylvania. He was totally independent so when he wanted to take off, he took off. I guess that paralleled the experience of growing up in a fishing family. And like kids that grow up on fishing boats, I learned early on that I could earn my own money.

I started fur trapping in elementary school. Trapping's like set netting—there's that anticipation of what you're going to have in the morning. You'd be out there at first light in that crisp November air, crunching across pastures of frosty grass, wading streams in your hip boots. By the time you caught the bus for school you'd had a little adventure and made some money.

My last season, before going off to college in 1968, I sold enough muskrats and other pelts to order a goose-down mummy bag from Eddie Bauer—the old Eddie Bauer that used to equip mountaineering expeditions. It's seen some hard use over the years but I've still got that bag, and I'm proud of it—paid for with fur, you know, all renewable.

At the time, we didn't think of it as "commercial fishing" but that's pretty close to what we were doing when another boy and I were catching snapping turtles. We'd set lines in farm ponds, and sell the snappers at a country auction, for soup.

You'd fasten one end of the line to the bank, bait the hook with a sunfish and whip it out in the pond. One morning, the ground was all trampled around the log I'd tied the line to and the hook was gone. I'm not proud to say it but we didn't hang around to verify whether it was stuck in the leg of one of the farmer's Holsteins.

During the summer, the real money was in landscaping, lawn maintenance. I'd cut grass all day, and at night I had my clients with outdoor lights. I'd mow until ten o'clock which, in hindsight, may not have been popular with the neighbors. But no one ever complained, I guess they didn't want to slow me down.

Dad was born in '07, so he was pretty well set up by the time I started driving. But the law remained, "If you want a car, you buy it."

I was soon driving to school in an old Jaguar roadster that leaked oil like an Exxon tanker. To me, there seems to be an unwritten law in the universe that the better you are at reading and writing, the worse you are at mechanics. Back then I didn't know a camshaft from a crankshaft, and stayed that mechanically ignorant right up until I started fishing and had to learn.

That was the main advantage that fishermen had on me. Commercial fishing, the whole thing boils down to the engine in that boat, and I've never known a fisherman that couldn't tear his motor down, rebuild it and tune it to perfection. They pick that up as kids, watching Daddy and messing about in their own little boats. Apart from that, I had a good start on the basic requirements—ambition, self-reliance, resourcefulness.

Improvising with what you have, nobody beats fishermen in that respect. Mike Voision, who was a big oyster dealer in Houma, called them "MacGyvers," after that guy on television. And really, when you're out on that boat and there's a problem, nobody's going to fix it for you, and it's got to be done with what's on hand.

I guess I got some of that from my dad. He had an old English shotgun, a double-barreled flintlock, and we used to play with it. The ramrod fit in under the barrels and it was partially wrapped with a piece of silver trim that was shaped kind of like a fish. It was pretty intricate, and after one of our war games it ended up missing. Instead of rushing off to a gunsmith, Dad took a silver dollar, put it on an anvil and started to whack it with a hammer. He beat it into a smooth, flat piece of thin plate, then trimmed and sculpted it until it fit perfectly.

Besides resourcefulness, determination's a big help. Old Winston Churchill once gave a commencement speech at Oxford. When his time came, he stood up, walked to the podium, said, "Don't give up!" went back and sat down.

That's what it takes to make it at anything but, with fishing, you get in these slumps, you can't catch anything, the bills are piling up, and you have a major breakdown. You wheel and deal to put it back together, go out and break something else. Insects, wind, cuts and bruises, poor seasons, physical exhaustion, regulations, you hang in there through all that and one day, jackpot!

Of course, you've also got to know a little bit about fish.

The fishing started with Dad. You're a kid, you see him tying flies, messing with all this neat gear, packing up and going off on trips, and that's what you want to do.

When Dad moved from the city in the late 1940s and built our home, the whole township was just one hundred-acre farm after another. Every farm had a pond and I had the run of most of them.

I learned to scout them, walking along the bank looking for largemouth bass. A lot of times, they'd take off before you could see them but they'd leave behind a muddy swirl that gave you a pretty good idea how big they were.

Dad wasn't a big bass fisherman but in the fall, he'd take us out for smallmouths. The Susquehanna's a lovely river, rocky with wooded islands and grass patches. We'd take a boat to an isolated stretch, beach it on an island and wade.

Trout were my father's main interest, and Pennsylvania's full of beautiful streams. We used fly rods, but not always flies. Dad wasn't a purist in that regard until later.

Early each spring we'd drive out to a press, where they made apple cider, and pick through this mountain of pommings, the squashed apples that they'd dumped out on a hillside the preceding autumn. You'd turn that half-frozen fermenting goop over and there'd be hundreds of worms. Dad called them "pommings worms" and I've never seen them anywhere else. They weren't big, but beige-colored, ringed with burgundy stripes. Pretty worms.

We'd get enough for the whole season, and keep them in wooden buckets in the basement. When we'd go fishing, we'd carry them in cylindrical bait cans that you wore on your belt and rotated to open.

We'd tie on a pair of small hooks, about two inches apart, bury one in each end of a worm and fish them like a nymph. Just flick them upstream and let them drift down naturally with the current.

In the swifter streams, you had to use a little weight. The split-shot came in tiny turquoise-colored tins, and on the bottom was printed, "Take a boy fishing."

LET THE GOOD TIMES ROLL

Home was just a few hours from the Atlantic Coast so we did a lot of saltwater fishing. We trolled little darts for shad down at Havre de Grace, where the Susquehanna flows into the Chesapeake. We'd also make trips for striped bass, trolling foot-long eels behind Dad's aluminum skiff. At the end of one fruitless trip, we anchored over a patch of eel grass, baited our hooks with bits of bloodworm and nabbed some little ones. They weren't big enough to keep but at least they were stripers.

Occasionally we'd charter a boat and fish for white marlin out of Ocean City, Maryland. Later, my brother got a 24-foot Mako offshore sport-fishing boat, and we started to get more serious.

It was an eight-hour drive to the Outer Banks of North Carolina. Dad had made that pilgrimage every autumn since the late '40s, in search of his Holy Grail—a channel bass over 50 pounds. That's what they call redfish on the East Coast, and 50 pounds is a big one. He came close, and made films of himself releasing 30- and 40-pounders back into the surf.

My oldest brother fishes really hard. I tell him he should have been the commercial fisherman. Once he got that boat, it was blue marlin, tuna, all that stuff out of Hatteras, sailfish off Florida, you know, just Fish! After I left home, he and Dad kept at it and got better and better.

Every Memorial Day, another pilgrimage would be to Delaware Bay, for the spring run of weakfish or "gray trout." That's where I first saw someone fishing with a net.

We were anchored up out there and this commercial skiff pulled up, no more than a hundred yards off. They ran out a stab net and stayed there, tied to it. An hour later, they hauled back. They were using a power roller, which raised the net high above the stern, and we could see they had a few nice trout, a few more bluefish, not exactly a swinging load.

ROBERT FRITCHEY GOLDEN MEADOW & LEEVILLE

It was interesting but no big deal. There was certainly no, "Oh my God, they're raping the sea! They're catching our fish with that indiscriminate entangling gill net!"

Of course, it happened to be one of those magic days. We were out in 14 feet of water yet these huge trout were up on top, feeding in syrupy-thick swirls around the boat. The weakfish we caught ranged up to 14 pounds and, laid side by side, they went all the way across that Mako.

The next time I saw someone fishing with a net was in Louisiana. I was in grad school—-still 100 percent sport fisherman—and fellow student Frank Cogswell and I were fishing down at Empire. It was cold, crystal clear after a Northwester and we were drifting over some deep borrow pits that had been dug out of the marsh for fill.

Eventually, we ended up by a set net, run straight out from the bank. All along it you could see the tails of the redfish, out of the water, waving forlornly. Of course, I didn't know it at the time but when you see fish like that, well, they're just the ones caught near the top of the net. Sure enough, we watched as a bent-up old boy in a slicker suit shoved a little flatboat off the bank, putt putted over and picked it up. Then you could see the fish.

But it didn't bother us. In that cold weather the fish had migrated to those deep holes and the trout had been ripping the surface into a froth all morning. We had about 75, most of which we probably ended up feeding to the cats.

We were sated, no envy, no emotion. Just, "Hey, check that out." If we hadn't caught a damned thing, would we have resented that old boy? I'd like to think that we were a little more sportsmanlike than these people today. After all, we were contented just to be out of the city for a while. We didn't *have* to catch fish.

LET THE GOOD TIMES ROLL

Of course, that was back in the mid-1970s, pre-CCA and before the outdoor writers became so greedy. We were hungry for hard information back then and we used to study *SaltWater Sportsman* magazine, all the books we could get our hands on, plus the outdoor columns in the dailies. You just didn't see the demeaning rhetoric that later became so prevalent, and if they did mention commercial fishermen, it was usually to relate what they'd learned from one!

By the '80s, though, after I'd gotten into fishing, I remember going home and Dad would mention "netters." He'd kind of growl it out in the same tone that some people apply to another "N" word. Of course, by then he was well into the Atlantic Salmon Federation, and nobody resents net fishermen more than those guys.

Atlantic salmon fishing's called the "sport of kings." And, by definition, it is, because European royalty have long pursued the species. The wilderness settings are also majestic, and the echoing roar of the powerful rivers seems to applaud the mighty angler. Then there's the fish, maybe 20 or 30 pounds, erupting from the dark water, time after time, flashing silver, living up to its name, *Salmo salar*—"The Leaper." Even the flies are regal, everything about it is, really, including the fees.

Dad used to tell me that, sitting around in a Scottish castle or a tent in Russia, after a day on the river, he'd sometimes feel a little insecure around some of the other guests. Not in fishing ability, but financially. A lot of these fellows operated on a level that you just didn't encounter around Harrisburg, PA.

Some of CCA's top dogs were salmon fishermen and after the Florida vote, the president of the federation wrote an editorial in its magazine titled, "The lesson from Florida: Ban gill nets forever." After praising the publisher of *Florida Sportsman* for his efforts, then listing other "successes" in Texas, California, Louisiana and Mississippi, he

ROBERT FRITCHEY GOLDEN MEADOW & LEEVILLE

closed with, "It's time we pushed gill nets irrevocably into the past, like thumbscrews, iron maidens, and other instruments of horror."

After years of reading that sort of crap, and hanging out with the entitled anglers, I could understand why Dad—easy going, generous, a great guy who could actually catch fish—had begun to come down on "netters." Fish passion, no one's immune.

Graduate school had brought me to Louisiana and after earning a Masters, the way was clear to keep on going, all expenses paid. But after so many years in school, penned up inside, topped off with a disappointing divorce, my true self came out: "I'm going fishing."

Frank and I had put together a rig, a vintage 14-foot Arkansas Traveler aluminum flatboat, a ten-year-old 18-horse Evinrude, and a rusty trailer. We raised the money doing drug studies at Tulane.

They tested an anti-coagulant on us one weekend, hit us up with a dose of coumarin, turned us into walking hemophiliacs and then were afraid to let us leave.

Once we had the rig, we started to explore the marsh which at first was scary as heck. Compared to the comforting trees and hills that I was used to, it was wide open out there. And back in the '70s, unlike now, you hardly saw anybody.

We fished down the river at first—Pointe a la Hache, Empire, Port Sulfur, Hopedale, Shell Beach. Then fell in love with the country down here in Lafourche Parish. Later, when I'd had enough of classrooms and cities, I bought my partner out and moved down into the marsh.

I arrived in May of 1980 and was still in the tent in December when Nolte Griffin, a hard-driving entrepreneur who owned the marina in Leeville, made me an offer. Nolte and his sons had done well selling

LET THE GOOD TIMES ROLL

fuel and ice to the fishermen, and had built a new raised camp across the canal from their old one, a little tar-papered trapper's shack that they simply abandoned. "You might have to set your tent up inside of it," said Nolte, "but you can stay in our old camp. It won't cost you anything, just keep an eye on things out there."

The camp was five miles by boat from Leeville. By living off the land like an old-time Cajun and with a cistern for water, kerosene and Coleman lanterns for light, a propane stove for cooking, and a wood-burner for heat, I kept expenses low enough to learn my new trade.

When I'd moved into the marsh, I had no intention of getting into commercial fishing. To the contrary, I was getting my first story published, in a national sport-fishing magazine. The piece was called "Gentlemen Prefer Reds," and there I am, Joe Cool sport, with my double-brimmed bonefishing hat, holding a "brace" of twelve-pounders, telling everyone how much I knew about catching redfish.

But when I started to support myself by selling them by the pound, well, it didn't take long to find out that I really didn't know anything.

The operative plan had been to camp out in the marsh, living off the fish I caught with my rods, until I felt competent enough to skip down to Belize and open a sport-fishing camp on one of those outside islands.

I'd spent some time in the tropics, related to my studies in tropical medicine, and had done some bonefishing around Contoy, an uninhabited island off the Yucatan. So the idea wasn't that farfetched. There were just a few holes that needed to be filled in first.

But by the time I'd become capable enough to pull it off, like so many others who'd come to Louisiana, with no intention of staying, I got stuck in the mud.

ROBERT FRITCHEY GOLDEN MEADOW & LEEVILLE

A game warden prompted me to buy my first net license. Like most of the agents in the field, he was an okay guy. He and his partner had already checked me out one night at my campsite so they recognized me when they pulled up to my boat.

I was by a weir, a type of dam they used to build in the marsh. The weirs slowed down the influx of salt water, and kept the water level up behind them on a falling tide. The reds also piled up behind the structures and I was about to get started on a bunch when the wardens showed up.

The agents could see what I was up to from the size of my ice chest. In those days, it wasn't exactly legal to sell fish that you caught with your sport license, but things were pretty loose. Still, the warden suggested that if I was going to keep at it, I should buy a gill-net license.

"A what?"

When I went up the bayou to buy my license, the May shrimp season was about to open and the place was packed. It was all Cajuns, I was a stranger, just looking and listening, and when my turn finally came, I asked the guy, "Do you sell gill-net licenses?"

"Unfortunately."

By that time, the boys from Florida had showed up and they used gill nets, but strictly for striking fish. They really knew what they were doing. But the locals picked up on the fact that you could stake those nets out and catch fish, and they were beginning to pop up everywhere.

By then, to survive, I'd begun to run some crab pots, which didn't require half the knowledge that fishing did. So I took my new license and went back to crabbing and rod-and-reeling, until late in the summer, when I made a trip with an alligator gar fisherman. He'd come down from the swamp country and was setnetting for them with flag nets—nets that didn't have a leadline, which allowed these primitive fish to get to the top to breathe. I worked with him awhile, then broke out on my own and started to earn a better living. Later, I apprenticed under an old Cajun boatbuilder and trammel netter, we custom-built "the perfect redfish boat," and I graduated to strike netting.

By the mid-1980s, I'd moved to terra firma in Golden Meadow, bought a house and a new truck. Everything I owned, my clothes, my food, my insurance, was paid for with redfish, with some trout, gar, drum, flounder and sheepshead thrown in—all renewable resources, right from the marsh.

Twenty years later, I take my sport-fishing buddies to the same places I used to work, and there they are, each pond with just as many fish as they always held. Environmentally, it doesn't get any better than that. And, after starting from zero, well, I was glowing, not to mention lean, tanned, and hard as a rock.

Back then, you worked hard and you got rewarded. I felt like I'd been shot from a cannon, like a new immigrant, like the Vietnamese: "You work hard, you buy Trans Am."

I tied up my boat by J&L Seafood. Jane and Lonnie Black were doing the same thing I was, but on a higher level. Jane was a real businesswoman.

She had a long German maiden name, blondish hair, and was originally from Ohio. While living in Florida she fell for Lonnie Black, who was from a fishing family out of Port Salerno. Lonnie's brother Ira had come over in the early 1970s and opened a little fish house in Leeville. After Jane and Lonnie bought him out, they started to build it up.

ROBERT FRITCHEY GOLDEN MEADOW & LEEVILLE

In the early days, you'd walk into the fish house to get paid and Jane had her office walled off from the rest of the shed with wooden fish crates. She had a cot in there and there'd always be a book on her pillow on accounting, marketing, business.

After the sportsmen took the trout and reds in Texas in '81, the Blacks opened a place over there. Jane bought the fish from us and trucked them over to Lonnie, who dealt them out of Houston for top dollar. But that was just one thing. They were sending our fish to New Orleans and all over the country.

Jane could move any kind of fish. She was buying reds, trout, all the traditional inshore stuff. Roe mullet, that was just starting and it was going to Taiwan. She had the offshore fleet selling her red snapper and grouper, most of which she sent to the Fulton Fish Market in New York. She was also developing brand-new markets for underutilized species that most of the other dealers didn't want to mess with.

Garfish, that was an old fishery in freshwater country, but Jane was the first person to buy them down on the coast, and they were thick. Gafftopsails, a hefty sea catfish, she started buying for 12 cents a pound. Now, that's not enough to make you target gafftops but if you'd had a poor day chasing trout on the beach you could strike a school on the way in and pay your expenses. And she could get a new little market going, that might grow to where she could pay us 50 cents. Things like that were insurance for down the road, to help pay the bills if they took the money fish. Of course, back in the '80s, nobody dreamed they'd take everything!

The Blacks even opened a restaurant, Jane's Catch, which was glassed in and overlooked our boats that were tied up on the bayou.

They worked non-stop, building their business. But, coming from Florida, which was about a century ahead of Louisiana in the amount of tourism and waterfront development, they'd seen seafood producers forced out of a lot of areas. And in the late 1980s, Texas banned the importation of wild-caught reds from other states, which hurt that part of their operation and gave the Blacks another peek into the future.

So Jane unloaded her place on the Griffins, the same family that had helped me out. Nolte's son-in-law Archie runs Griffin's Seafood now, and he deals mostly in offshore fish like snapper, grouper, tuna, and king mackerel—stuff managed by the federal government, not the state of Louisiana.

The Blacks eventually hooked up with a New York outfit that had a reef-fish dock out in the western part of the state. So, instead of continuing to invest in their own business, they just started to buy fish for them, which was too bad—if the sportsmen hadn't gotten carried away, there's no telling what Jane and Lonnie would have built by now.

Throw in all the other fishing families along the coast who were building their businesses in the same way and it's scary what our fisheries—and our wetlands—would be worth today.

We all wondered about Jane when she ran for state representative. She had no sooner opened her place, and here's this stranger—from Florida and a woman to boot—running for office, amongst all these Cajuns. People dismissed her campaign and, no, she didn't make it. But her picture was on every pole and when people would ask, "Who the heck is Jane Black?" the answer was always the same: "She's that lady who just opened up that fish house in Leeville. J&L Seafood."

Yes, she was running to put some pressure on our incumbent representative, who was beginning to lean toward the sportsmen. And in the four-way race, he was unseated. But at the same time, she was marketing her business, which was pretty shrewd and resourceful.

When the sportsmen started to threaten our trout fishery, Jane saw the need for an organization, to protect us and, of course, her own business. She patterned it after the OFF, the Organized Fishermen of Florida, and called it the Organization of Louisiana Fishermen.

Jane ran the OLF with a couple of the fishermen's wives. When they'd be working the Legislature, she'd call a board meeting. We'd all be sitting there, fried from a day on the water, trying to stay with her as she briefed us on the serpentine progress of that session's anti-fishing bills.

"Don't you think this is how we should handle it?"

"Yeah, Jane. That sounds good."

Eventually we hooked up with the Catholic Church. If the Catholics can't back up fishermen, who can?

Actually it was an affiliate of the church, the Catholic Social Services, and they started to give us grant money through their Campaign for Human Development. They wanted results though, and told Jane, "We'll give you the money but only if you agree to be trained by a community organizer."

Enter Wade Rathke, chief organizer at ACORN. The national Association of Community Organizations for Reform Now builds community organizations of low and moderate-income families, with the philosophy that social change should come from the bottom up, rather than the top down. Wade had started ACORN, and he helped Jane build the OLF into a powerhouse of a group, with 350 members. That's 350 net fishermen, reef fishermen and shrimpers, each one a bullheaded captain and business owner.

We worked up to getting our own building, which we shared with a nun. Sister Melinda helped folks get their GEDs, and during

LET THE GOOD TIMES ROLL

"Friday school" she taught people from the bayou how to read.

When a major landowner locked up a lot of marsh, the OLF sued him on the principle that nobody owned tidal waters. The lawsuit dragged on for years, until the sports took the reds and the other fish, and it didn't much matter anymore.

In an election year, the OLF held forums. We'd write out questions for the candidates and, in front of a packed hall, ask them what they thought about various issues. And record every word for future reference.

The OLF was a growing force in the community, a positive force. But that's all gone now of course—you can't have a trade organization without a trade. Running the group was a tremendous strain and with our government dimming her prospects, the costs outweighed the benefits for prime mover Jane. But before she moved on and the group went down the tubes, I was fortunate enough to have hooked up with the Coalition to Restore Coastal Louisiana.

In the late 1980s, when word first started going around that GCCA was going to make a run for Louisiana's redfish, fishermen down the bayou said, "They can't do that." Even when they set the final quota, no one really believed that they'd shut down the centuries-old commercial fishery, until its final weeks.

When you can fish year round, you pick your days, according to the weather. But with an impending closure, and the wolves at the door, you go every day, wind or rain, and that's what I did right up to January 15, 1989, the last day we fished the reds. My deckhand Frank McCain and I were hauling in at least as many fish from our regular spots as we always had, maybe more. That experience, on top of everything else I'd learned, gave me a leg up, after I let everything go and moved to the French Quarter, to try to figure out what had happened.

While I was there, Graham Wisner, one of the coalition's board members, threw a benefit at his place, and I went. The OLF was a charter member of the coalition—Rob Gorman at the Catholic Social Services had hooked us up with them. Rob was one of the coalition's founders, along with the other prime movers like Oliver Houck, at Tulane Environmental Law, Jim Tripp at the Environmental Defense Fund in New York, and a few others.

They'd first put this coalition of organizations together to demonstrate that a wide variety of interests had a stake in saving the marsh. For the first couple of years, the OLF was just a warm body, we paid our dues and that was it. Jane never attended the coalition's meetings—she had her hands full with the fish politics, beating back regulations, to keep the industry going. But the marsh was my thing so, after that gathering in the Quarter, I got involved with the movement to save the wetlands.

When a board meeting would come up, I'd call Jane, she'd raid petty cash, send me 25 bucks, and I'd catch an early morning bus up to Baton Rouge. Later I'd catch rides with fellow board members and we'd educate each other on the way.

Associating with these folks was inspiring because the coalition's board held the crème de la crème of the state's environmentalists, and squeaky clean representatives from churches, business, science, and sport and commercial fishing. Everyone was a friend, even the sportsmen with the Louisiana Wildlife Federation, who'd always been at the forefront in saving fish and wildlife habitat.

GCCA, of course, had bigger fish to fry: When asked to join, the single-issue group dismissed the offer: "We don't join coalitions." That kind of opened up the eyes of some of the environmental movers and shakers.

LET THE GOOD TIMES ROLL

Later, true to form, the group claimed in its literature, *"We're* the group that got the resolution passed to save Louisiana's coast."

Initially, the biggest obstacle to saving the marsh was the lack of a sustainable source of funding. So the coalition produced a report, "Here Today, Gone Tomorrow," that outlined the problems and their solutions, and used that as a tool to convince the Legislature to allow the public to vote on the issue. In 1989, voters overwhelmingly approved the coalition's trust-fund measure and the state began to set aside up to $26 million a year for coastal restoration. That's not much but it was leveraged big-time with federal money.

The restoration of Louisiana's coast was probably the biggest public works project in the nation's history. Actually, it involved the integration of many large-scale projects that had to be planned, nominated, prioritized, coordinated by a variety of state and federal agencies and, eventually, built. The coalition worked to streamline the process but it would always be necessarily ponderous.

Typically, from the time they dreamt up a project, until they dug the first shovel of mud, you were looking at twenty years, minimum. In that time, we'd lose at least 500 more square miles of marsh, while the project itself might bring back a hundred. It didn't take long to figure out that, at that rate, we on the coast were toast.

Meanwhile, I was working on "Wetland Riders," a book I'd hoped would get some folks behind us when it came time to reopen the reds. As part of the deal, I'd proposed either an annual habitat stamp on commercial fishermen, or a severance tax on our products, to set up a streamlined wetland restoration/fisheries enhancement effort within the state's Department of Wildlife and Fisheries. That, too, could be leveraged with grants from NOAA and a variety of other sources. We

ROBERT FRITCHEY GOLDEN MEADOW & LEEVILLE

even figured we'd shame the "sports" and oil companies into contributing.

Our idea was that when you'd order a wild-caught redfish or flounder in a New Orleans restaurant, you'd not only be tangibly linked to the marsh, but also paying to run the equipment that would, at least, slow down its loss until the really big projects came on line. We were even looking at "infomats," arty place mats that showed reds and other seafood species in the marsh, to subliminally make the link to diners.

But when the book came out, in October '94, here I was, promoting us fishermen as part of the solution when everyone else was portraying us as the problem. Just a month before the vote in Florida, I was immediately drawn into that fight and, later on, most of the others around the country. When the dust settled, we were out of business here.

Of course, 1994 was also the first year since we'd lost our fish that state biologists came out and said we could safely have a quota of reds. So the timing was perfect but instead of celebrating the Seafood Promotion and Marketing Board's planned "Return of the Reds," it was, "Nyahh nyahh nyahh, you can't have any!" Not quite the same message.

The OLF was going to promote the action fund legislatively but after Jane sold her business and moved on, the OLF collapsed.

Non-profits are like that, they need a prime mover or two to get going and, once over the hump, they're set up to where they can hire a staff and keep it going. The OLF had a good run but didn't make it over the hump.

I figured it was all over, when out of the blue, something happened that lifted our hopes, for a while. The wife of a Lafitte fisherman had put together a little fishing group over there. While working on a project at Tulane's Environmental Law Clinic she came up with the idea for a new coalition.

AFFERM stood for the Alliance for Fair and Equitable Resource Management, which you have to admit was a pretty good name. As chance had it, the initial meeting of the group coincided with the first day the net ban took effect, March 1, 1997. That was when the realization really struck home: "We're in South Louisiana and not one soul is out there netting fish to eat." It was surreal, and profoundly sad. Things had clearly gone too far, and we aimed to do something about it.

Tracy Kuhns, who'd started it, either she or her children had some health problems, which she attributed to the pollution from having lived in Texas, the only state more toxic than Louisiana. Inspired by Diane Wilson, the Texas shrimper who reduced pollution policy to a simple "zero discharge," she brought in Mary Lee Orr and her Louisiana Environmental Action Network (LEAN) to work toward the same goal.

I was still on the board of the coalition, so the Coalition to Restore Coastal Louisiana came in. The Lake Pontchartrain Basin Foundation had worked with Cliff Glockner and other fishermen, so they came in. Daryl Malek-Wiley's Environmental PAC for Louisiana, virtually all of the state's environmentalists came in with us.

The driving force was fishermen, though, and we had groups from all across the state: Kerry LeBauve's Committee for the Future of Louisiana's Traditional Commercial Fishers, Bayou Segnette Community and Boaters, Delta Commercial Fisheries Association, Lake Catherine Fishermen, Louisiana Commercial Fishermen's

ROBERT FRITCHEY GOLDEN MEADOW & LEEVILLE

Association of Dulac, Pointe Aux Chênes Tribe, Terrebonne Fishermen's Association. And the ones that didn't join were watching.

Besides us fishermen, who needed access to the resource, Frank Brigtsen came in, and he was our liaison with all the other chefs who wanted to keep serving a variety of locally caught wild fish.

Frank had invited me to speak with him at a national convention of chefs, where we alerted them to what was happening and encouraged them to get behind their own fishermen. Through the day, Frank and I would shoot the bull with chefs he knew from California, New York, all over. I'd ask them what they were using and the answer was always the same: "Farm-raised Atlantic salmon and farm-raised catfish." When it was all over, I asked Frank why he was helping us out. "So *I* don't have to serve farm-raised salmon and catfish."

Of course it was about more than having a top-quality product for Frank and other home-grown chefs like Paul Prudhomme. Great cooking does begin with the main ingredient but they also knew a lot of the people who were involved, and what it all meant to South Louisiana's culture and heritage.

AFFERM directly addressed the Holy Trinity—water quality, habitat, and the equitable allocation of our resources. The interests of the local environmentalists and the fishermen naturally coincided, but AFFERM formalized the relationship, to everyone's advantage. In other words, if our local enviros were going to promote a bill to clean up the water, we'd be right beside them, "This is important to us economically because we can't market polluted oysters or tainted fish, and the fish can't reproduce in toxic water."

Our whole thrust was "sustainable coastal communities," and we had to be all-inclusive. You needed the seafood industry, but you also needed the bait shops, the motels for the sports, the charter fishermen, eco-tourism, the full complement of interests that depended on the wild resources. Until the GCCA came to town, everyone had merely shared the resource and gotten along without thinking of excluding anybody. We aimed to get back to that point.

It was hard to argue with and to make our case we started on a report that was patterned on the coalition's "Here Today, Gone Tomorrow." I'd write the first draft, we'd meet and edit it collectively to eliminate any bias.

We worked on that for a year, while we brainstormed, reached out to new people, made sure it was real. Once again, the Catholic Social Services offered its help, promising to bankroll our executive director for a couple of years. They also offered to help organize the hearings that we were about to hold across the coast.

People on the coast were fed up with hearing about well-funded "studies" that were rarely followed up with action. When they heard that we intended to kick it up a notch, set up a lean and mean task force and get machinery in the marsh tomorrow, well, it seemed like a done deal to me. So I started lining up projects.

The city of New Orleans owned about 35,000 acres of marshland in Lafourche Parish. The land had been donated by Edward Wisner, and was held in a trust, with the city, Tulane and other entities sharing the oil revenue it produced. They also leased sites for camps, and I had one. I'd worked that marsh for years and used to know every clump of grass. So I took the attorney who ran the operation out there for a ride, told her what we were going to do and suggested we use the Wisner tract for a demonstration.

It sounded good to her. Married to a prominent coastal geologist,

ROBERT FRITCHEY GOLDEN MEADOW & LEEVILLE

When redfish were worth money, landowners routinely hired excavator buggies to maintain their marshland. Though it represented a first line of defense against wetland loss, the practice has largely been abandoned.

she was well aware that the trust's land was washing away. She could obtain a general permit to get started, she said, and even offered to throw in some money.

It wouldn't have taken much to get started. For $300,000, you could hire an excavator buggy, a sort of amphibious steam shovel, to work out there full time for a year. "Marsh maintenance" wasn't rocket science but like ACORN's notion of getting things accomplished from the bottom up, the push had to come from the grassroots, the folks who lived and worked in the wetlands.

LET THE GOOD TIMES ROLL

But, before we could take our case to the public, our position began to erode even faster than the marsh. Our prime mover developed problems, couldn't go on, and that was it for AFFERM. The idea that had shown so much promise didn't quite make it over the hump. It probably seems crazy to folks who have steady jobs, steady incomes, but it took a day's work just to line up everyone for a meeting. And after they took the nets, we were hanging by a thread.

As the full effects of the net ban sank in, saving the marsh took a back seat to survival.

While working on my book in the Quarter, I was also rebuilding my boat, getting ready to move back down the bayou. Even if we didn't get the reds back, at that time you could still net trout and every other kind of fish. But after the book was out and it was time to go, the only thing left was mullet and I couldn't even qualify for that license because I hadn't earned at least 50 percent of my income from the sale of seafood, within the stipulated years.

Still, a windfall enabled me to buy back the same house in Golden Meadow that I'd lost with the redfish in the '80s. This time it came with a tenant, so I'd have a little income to help me along. After five years cooped up in a Decatur Street slave quarter, I relished the thought of moving back into the marsh. And with a camp out there, that's what I did.

After they took the nets, if you didn't get out of the business, like Cliff Glockner did, you got into either crabs, shrimp or oysters. With the stinking bait and stinking thieves, I'd gotten enough of crabbing in my earlier days, so started shrimping and oystering.

Entry level oystering, you put on chest waders, drag around a pirogue, and pick them up one by one. It's hard to take over the world

ROBERT FRITCHEY GOLDEN MEADOW & LEEVILLE

Netting red drum from a shallow-running 18-foot "mudboat."
(Brian Gauvin)

LET THE GOOD TIMES ROLL

ROBERT FRITCHEY GOLDEN MEADOW & LEEVILLE

picking up oysters by hand.

Because you look like a raccoon feeling around in the water with its paws, oystering by hand is called "cooning." It's done in the wintertime because the tides are at their lowest and the oysters are at their best. During the summer, I trawled the marsh for shrimp. After a couple seasons, it sank in that I needed more boat.

Netting fish from my 18-footer, at a buck a pound, you could put $500 worth of trout or reds in your box, get up on top and run home. But $500 worth of oysters, at $12 a sack, that weighed more than two tons.

Since the shrimpers were trading their high-maintenance wooden boats for fiberglass hulls, there were a lot of old Mediterranean-style cypress luggers on the market. Like Kerry LeBauve, I ended up with one.

The old lugger came rigged for shrimp, with a set of skimmer nets. I fought with that for two seasons before yanking off the shrimp gear and rigging the boat to dredge oysters. I hung with that for a couple of years, and it was good to be earning a living again. But it wasn't fishing.

Fishing, you cover a lot of territory, hunting. You've got to figure out where they're going to be and when, and if you're right, you might find yourself jumping up and down with your deckhand, laughing, passing high fives. There's none of that in oystering.

The dredge dumps a mess of oysters, shells and old beer cans on a metal table, you stand there with your little tomahawk, cull out the good oysters, shove the shells back overboard and dump the dredge again. Standing out there on deck, bustin' rocks, I had to remind

[Facing page] At entry level in the oyster industry, after the net ban.
(Brian Gauvin)

myself, "I didn't kill anybody!"

I'd first gotten into this business because I loved to fish but here I was, grinding away. I could do other things, however. I wasn't as trapped as the boys who were born into fishing, quit school in the eighth grade to support their family and, decades later, were fishermen to the bone.

Twenty-five years after I moved down to Leeville and started out with nothing but a rod and a reel, I sold the oyster boat and started to write this book. It had been a heck of a ride, and I'm way better off for it. My only regret is that if someone wanted to follow in my footsteps today, they couldn't.

Appendix

TOOLS OF THE TRADE

TABLES 1–5

BIBLIOGRAPHY

TOOLS OF THE TRADE

A GLOSSARY OF SOME FISHING NETS

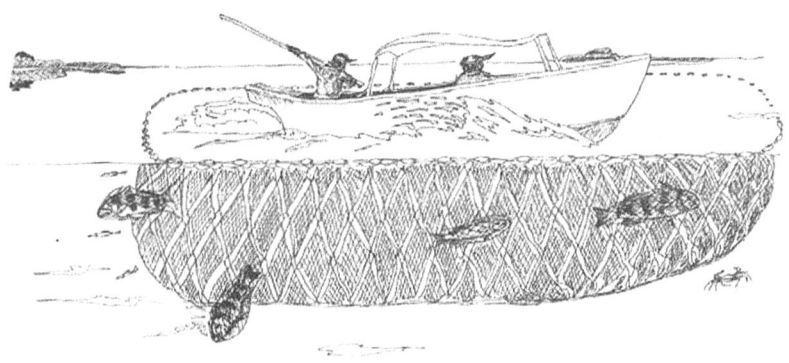

TRAMMEL NET

The trammel net is actually three nets in one. Sandwiched between the two outer "walls," the inner net is fine-meshed and hung with plenty of slack. The walls are nets with extremely large meshes, up to 18 or 20 inches. When a small fish hits the trammel net, it may be gilled in the inner net; when a larger fish hits, it forces the small-meshed inner net to bulge through a large mesh on the opposite side, neatly bagging itself in a pouch.

The trammel net's construction enables fishermen to catch deep-bodied species lacking prominent gill covers—such as pompano, flounder and the freshwater buffalo—as well as a wider size-range of fish than they can with gill nets.

LET THE GOOD TIMES ROLL

GILL NET

Gill nets trap fish as they try to swim through their webbing. If the fish is the correct size, its head will pass through a mesh but its body will be too large to follow. Unable to go forward, and prevented by its gill covers from backing out, the fish is held until removed by the fisherman.

Obviously, the larger the fish desired, the larger the mesh size must be. The selectivity of the gill net can prove frustrating for fishermen as smaller, yet marketable, fish pass through the webbing. Fish that are too large to enter the meshes often back off and escape, as well. Still, the gill net's light weight and ease of handling and clearing more than compensate for the fish lost due to its discriminating construction.

The gill net's selectivity is also useful for fishery managers, who can mandate minimum mesh sizes that won't catch fish until they've grown large enough to have spawned at least once, thus ensuring continuous production.

Gill nets may be fished actively or passively. A runaround gill net, or "strike net," as it's called on the Gulf Coast, is fished actively: When the fisherman sights a school of fish, he throws over a weight or buoy that's attached to the net which is then pulled into the water as he circles the fish. After they hit the net, the fisherman hauls it back onboard.

A "set net" is fished passively by anchoring it in a promising location, like a spider's web. Fishermen check their gear periodically to clear it of fish.

TOOLS OF THE TRADE

Gill Net (Strike Net)

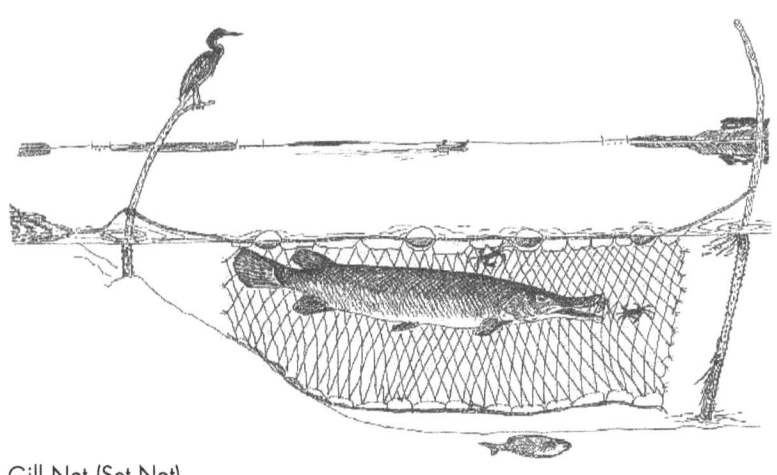

Gill Net (Set Net)

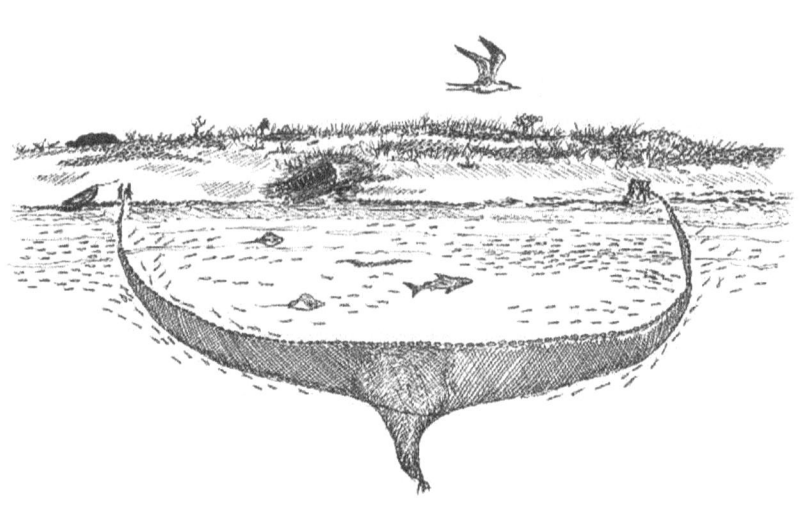

HAUL SEINE

Unlike gill and trammel nets, which entangle fish in their webbing, the haul seine surrounds the catch with a wall of small-meshed webbing that is constructed of twine coarse and durable enough to withstand the wear and tear this type of gear must endure.

After the seine is set around a school of fish, it is slowly dragged back to the boat or shore. As it is "hardened up," the catch is gradually forced into a bag of webbing that's sewn into the net. Once it's confined within the bag, the catch can be scooped out with a dip net or gaffed individually; either method is more efficient than clearing single entangled fish from a gill or trammel net.

TOOLS OF THE TRADE

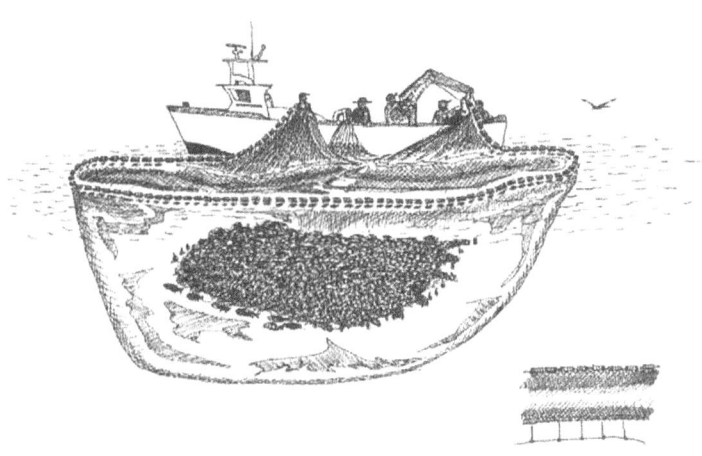

PURSE SEINE

The purse seine is a very large net used primarily to capture schools of fish found near the surface of deeper waters. Unlike trammel nets, haul seines and gill nets—the leadlines of which usually sink to the bottom to prevent fish from escaping beneath—the purse seine floats like a deep curtain of webbing. To prevent the catch from escaping by diving deeper, the net is "pursed" by heaving on a cable that's run through rings that are attached to its leadline. The net is then hauled in by power blocks or a net drum until the fish are herded to one end. They can then be dipped out with nets or suctioned aboard with fish pumps.

Tuna are harvested in the Pacific Ocean with purse seines that may weigh 50 tons. Smaller seines used to catch mackerel in the Atlantic may weigh 15 tons. The nets are widely used in Alaska's huge salmon fishery but on the Gulf Coast use of this high-volume gear is limited mostly to the menhaden fishery.

TABLE 1

LOUISIANA **COMMERCIAL** AND **RECREATIONAL** SPOTTED SEATROUT LANDINGS IN POUNDS 1990-2014

YEAR	COMMERCIAL[1]	RECREATIONAL
1990	648,645	2,679,167
1991	1,220,231	7,549,531
1992	971,481	6,382,263
1993	1,138,070	5,638,132
1994	1,023,687	6,705,959
1995	658,084	7,574,410
1996	774,474	7,594,638
1997	549,505	7,217,470
1998[2]	111,979	5,434,523
1999	76,261	7,802,459
2000	40,283	11,328,992
2001	109,468	9,555,273
2002	71,547	6,260,548
2003	19,401	8,221,350
2004	19,875	8,638,860
2005	16,948	7,676,539
2006	2,042	11,772,927
2007	11,050	9,775,855
2008	11,619	12,531,289
2009	999	11,195,495
2010	—[3]	8,442,029
2011	—[3]	12,951,159
2012	98	11,283,766
2013	1,246	9,381,579
2014	1,124	NA[4]
TOTAL:	7,478,117	203,594,213

[1] Commercial fishery capped by quota of 1 million pounds
[2] First full year commercial harvest limited to rod and reel only
[3] No landings reported
[4] Not available from NMFS

TABLE 2

COMMERCIAL LANDINGS OF GULF OF MEXICO RED DRUM IN POUNDS 1887-2014[1]

YEAR	Florida[2]	Alabama	Mississippi	Louisiana	Texas	TOTAL
1887	NA[3]	NA	141,000	289,000	1,005,000	1,435,000
1888	55,000	NA	165,000	288,000	944,000	1,452,000
1889	391,000	64,000	185,000	314,000	1,063,000	2,017,000
1890	458,000	54,000	201,000	339,000	1,108,000	2,160,000
1897	236,000	213,000	199,000	465,000	1,144,000	2,257,000
1902	1,104,000	70,000	93,000	442,000	898,000	2,607,000
1908	608,000	151,000[4]	244,000[4]	716,000[4]	1,309,000[4]	3,028,000[4]
1918	995,000[4]	23,000	116,000	566,000	1,337,000	3,037,000[4]
1923	1,398,000	15,000	177,000	665,000	878,000	3,133,000
1927	776,000	55,000	237,000	556,000	1,248,000	2,872,000
1928	889,000	49,000	208,000	434,000	1,030,000	2,610,000
1929	992,000	105,000	29,000	445,000	934,000	2,605,000
1930	937,000	104,000	122,000	335,000	873,000	2,371,000
1931	934,000	62,000	100,000	369,000	864,000	2,329,000
1932	719,000	44,000	75,000	282,000	825,000	1,945,000
1934	873,000	65,000	73,000	492,000	1,579,000	3,082,000
1936	927,000	34,000	88,000	347,000	956,000	2,352,000
1937	948,000	67,000	123,000	450,000	954,000	2,542,000
1938	1,012,000	32,000	106,000	522,000	860,000	2,532,000
1939	908,000	31,000	165,000	694,000	470,000	2,268,000
1940	647,000	27,000	55,000	183,000	265,000	1,177,000
1945	1,294,000	260,000	66,000	596,000	1,297,000	3,513,000
1950	941,700	16,000	51,600	455,400	567,200	2,031,900
1951	919,300	43,600	31,400	383,500	237,500	1,615,300
1952	646,500	55,700	41,300	327,700	249,500	1,320,700

TABLE 2, CONTINUED

COMMERCIAL LANDINGS OF GULF OF MEXICO RED DRUM IN POUNDS 1887-2014[1]

YEAR	Florida[2]	Alabama	Mississippi	Louisiana	Texas	TOTAL
1953	525,900	46,300	61,700	272,500	511,000	1,417,400
1954	752,000	18,700	60,700	270,500	721,100	1,823,000
1955	754,500	18,800	56,800	344,200	493,900	1,668,200
1956	762,600	49,900	71,300	407,000	639,600	1,930,400
1957	666,800	10,500	53,600	353,300	504,500	1,588,700
1958	626,900	19,500	65,000	487,500	599,000	1,797,900
1959	692,400	17,600	71,400	488,100	963,100	2,232,600
1960	817,000	9,000	38,900	428,000	704,900	1,997,800
1961	847,500	24,300	52,900	666,000	617,500	2,208,200
1962	1,307,300	12,900	76,000	567,200	699,400	2,662,800
1963	968,000	20,400	59,000	465,600	685,600	2,198,600
1964	699,200	19,300	50,100	311,700	446,900	1,527,200
1965	801,000	3,700	32,700	471,200	532,500	1,841,100
1966	645,000	6,100	36,900	531,400	797,400	2,016,800
1967	495,500	9,200	95,800	653,900	767,500	2,021,900
1968	707,200	16,400	214,600	740,000	924,900	2,603,100
1969	586,200	51,300	99,600	782,100	1,083,300	2,602,500
1970	667,500	35,200	70,300	789,200	1,586,200	3,148,400
1971	708,200	31,700	58,800	723,700	1,990,700	3,513,100
1972	843,400	77,000	55,700	889,000	1,467,800	3,332,900
1973	954,000	172,000	85,700	1,183,500	1,677,500	4,072,700
1974	1,191,200	119,700	88,600	1,436,100	1,921,500	4,757,100
1975	759,300	73,700	71,500	1,362,300	2,120,400	4,387,200
1976	904,820	66,600	95,200	2,212,500	2,029,400	5,308,520
1977	845,000	65,400	155,300	1,435,500	950,800	3,452,000

TABLE 2, CONTINUED

COMMERCIAL LANDINGS OF GULF OF MEXICO RED DRUM IN POUNDS 1887-2014[1]

YEAR	Florida[2]	Alabama	Mississippi	Louisiana	Texas	TOTAL
1978	899,082	86,360	658,000	1,218,797	861,102	3,723,341
1979	744,655	84,960	194,380	1,056,697	690,078	2,770,770
1980	816,730	52,530	20,430	724,777	1,114,332	2,728,799
1981	1,131,047	38,291	66,955	898,585	613,056	2,747,934
1982	860,936	69,137	40,600	1,454,503	0	2,425,176
1983	803,669	360,547	24,200	1,938,615	0	3,127,031
1984	848,614	853,536	23,660	2,608,383	0	4,334,193
1985	538,586	2,843,151	27,423	2,933,573	0	6,342,733
1986	878,371	5,304,691[5]	126,352	7,817,694[5]	0	14,127,108
1987	251,854	14,276	53,059	4,571,177	0	4,890,366
1988	4,583	785	41,109	245,365	0	291,842
1989	0	1,860	139,775	24,811	0	166,446
1990	0	0	5,166	0	0	5,166
1991	1,000	0	22,143	0	0	23,143
1992	0	0	62,551	0	0	62,551
1993	0	0	83,704	1,884	0	85,588
1994	0	0	40,246	2,957	0	43,203
1995	0	0	24,110	0	0	24,110
1996	0	0	30,363	1,925	0	32,288
1997	0	0	23,633	0	0	23,633
1998	0	0	30,798	4,769	0	35,567
1999	0	0	40,202	0	0	40,202
2000	0	0	8,084	0	0	38,084
2001	0	0	22,695	0	0	22,695
2002	0	0	17,863	0	0	17,863

TABLE 2, CONTINUED

COMMERCIAL LANDINGS OF GULF OF MEXICO RED DRUM IN POUNDS 1887-2014[1]

YEAR	Florida[2]	Alabama	Mississippi	Louisiana	Texas	TOTAL
2003	0	0	22,441	0	0	22,441
2004	0	0	18,457	0	0	18,457
2005	0	0	30,141	0	0	30,141
2006	0	0	22,192	0	0	22,192
2007	0	0	22,664	0	0	22,664
2008	0	0	28,011	0	0	28,011
2009	0	0	32,027	0	0	32,027
2010	0	0	6,444	0	0	36,444
2011	0	0	8,359	0	0	28,359
2012	0	0	4,797	0	0	34,797
2013	0	0	36,516	0	0	36,516
2014	0	0	52,258[6]	0	0	52,258
TOTALS	46,916,047	12,345,624	7,292,208	54,732,112	51,610,168	172,896,159

[1] From U.S. Bureau of Fisheries and NMFS Landings Data. Several years omitted between 1887 and 1950.
[2] West coast only
[3] Not available
[4] Includes some black drum
[5] Elevated landings in mid-1980s comprised largely of red drum caught in federal waters
[6] Yearly commercial quota increased from 35,000 to 60,000 pounds

TABLE 3

RECREATIONAL LANDINGS OF GULF OF MEXICO RED DRUM IN POUNDS 1981-2013[1]

YEAR	Florida[2]	Alabama	Mississippi	Louisiana	Texas	TOTAL[3]
1981	1,636,811	74,630	303,550	1,111,640	NA[4]	3,126,631
1982	2,418,323	201,290	366,528	3,428,716	NA	6,414,857
1983	2,428,957	41,208	583,744	3,824,079	NA	6,877,988
1984	3,457,050	179,910	400,610	2,841,326	NA	6,878,896
1985	1,961,065	66,105	214,637	3,193,275	NA	5,435,082
1986	1,662,339	202,617	213,521	3,110,958	NA	5,189,435
1987	565,486	103,679	279,779	3,620,363	NA	4,569,307
1988	143,788	111,337	137,412	2,539,519	NA	2,932,056
1989	1,232,421	33,292	95,497	4,379,744	NA	5,740,954
1990	759,443	107,588	171,446	3,014,515	NA	4,052,992
1991	1,085,181	86,319	190,867	3,999,416	NA	5,361,783
1992	1,642,032	163,926	160,931	5,833,875	NA	7,800,764
1993	906,427	171,601	207,568	7,424,691	NA	8,710,287
1994	1,060,865	255,253	298,590	5,913,250	NA	7,527,958
1995	1,105,093	376,366	567,380	10,046,848	NA	12,095,687
1996	1,285,187	218,656	981,746	9,358,599	NA	11,844,188
1997	1,411,832	518,320	792,601	9,481,006	NA	12,203,759
1998	1,268,005	446,386	474,590	6,333,248	NA	8,522,229
1999	1,016,914	638,743	326,337	7,266,304	NA	9,248,298
2000	1,721,290	408,352	417,717	11,693,763	NA	14,241,122
2001	1,296,830	1,111,234	484,743	10,526,594	NA	13,419,401
2002	1,232,760	706,523	507,927	9,128,555	NA	11,575,765
2003	1,593,624	773,780	500,141	10,245,642	NA	13,113,187
2004	1,262,957	1,056,485	522,641	11,513,227	NA	14,355,310
2005	2,050,084	825,885	339,318	7,691,630	NA	10,906,917

TABLE 3, CONTINUED

RECREATIONAL LANDINGS OF GULF OF MEXICO **RED DRUM** IN POUNDS 1981-2013[1]

YEAR	Florida[2]	Alabama	Mississippi	Louisiana	Texas	TOTAL[3]
2006	1,528,885	727,487	419,830	8,815,994	NA	11,492,196
2007	1,576,165	878,621	403,073	10,349,548	NA	13,207,407
2008	1,943,973	401,752	394,285	11,679,490	NA	14,419,500
2009	965,805	360,046	473,621	9,967,545	NA	11,767,017
2010	1,023,684	938,703	351,670	11,321,576	NA	13,635,633
2011	1,105,421	657,296	567,422	13,004,362	NA	15,334,501
2012	1,703,468	964,942	786,437	8,509,393	NA	11,964,240
2013	1,596,076	986,078	1,215,406	13,594,395	NA	17,391,955
TOTALS	47,648,241	14,794,410	14,151,565	244,763,086	NA	321,357,302

[1] From National Marine Fisheries Service landings data
[2] West coast only
[3] Excluding landings from Texas
[4] Not available from NMFS

TABLE 4

RECREATIONAL LANDINGS OF
GULF OF MEXICO **RED DRUM**
IN NUMBERS OF FISH 1979-2013[1]

YEAR	Florida[2]	Alabama	Mississippi	Louisiana	Texas	TOTAL[3]
1979	453,000	13,000	108,000	2,455,000	1,051,000	4,080,000
1980	555,000	27,000	177,000	1,705,000	940,000	3,404,000
1981	722,541	46,097	133,231	510,594	571,000	1,983,463
1982	624,377	3,747	135,710	1,695,906	568,000	3,077,740
1983	1,134,089	11,924	124,143	2,611,647	454,000	4,335,803
1984	1,340,213	23,254	65,973	1,427,062	530,000	3,386,502
1985	426,567	27,487	26,331	1,465,041	276,000	2,257,426
1986	614,820	59,069	115,343	1,625,984	330,000	2,745,216
1987	138,891	34,689	89,901	1,503,680	390,000	2,157,161
1988	28,521	16,321	49,766	812,776	287,000	1,194,384
1989	234,188	8,830	30,573	616,604	213,000	1,026,517
1991	279,048	24,329	34,804	872,713	211,000	1,632,894
1992	388,292	58,131	29,241	1,767,938	259,000	2,502,602
1993	197,195	52,252	32,013	1,913,832	451,000	2,646,292
1994	234,192	44,121	55,083	1,382,072	359,000	2,074,468
1995	259,084	74,409	81,965	2,449,022	335,000	3,199,480
1996	295,405	28,335	88,833	2,082,397	340,000	2,834,970
1997	329,296	50,761	79,663	1,830,537	348,000	2,638,257
1998	274,167	79,787	63,764	1,427,151	NA[4]	1,844,869
1999	228,794	85,244	56,137	1,763,495	NA	2,133,670
2000	377,151	58,269	56,265	2,774,046	NA	3,265,731
2001	266,329	136,274	60,406	2,652,282	NA	3,115,291
2002	291,894	83,716	60,370	2,041,827	NA	2,477,807
2003	365,170	114,013	49,996	2,143,424	NA	2,672,603
2004	320,999	117,600	83,120	2,418,339	NA	2,940,058

TABLE 4, CONTINUED

RECREATIONAL LANDINGS OF GULF OF MEXICO **RED DRUM** IN NUMBERS OF FISH 1979-2013[1]

YEAR	Florida[2]	Alabama	Mississippi	Louisiana	Texas	TOTAL[3]
2005	501,367	153,822	35,422	1,626,356	NA	2,316,967
2006	376,105	100,486	57,851	1,827,969	NA	2,362,411
2007	411,598	83,916	42,914	2,307,846	NA	2,846,274
2008	456,719	88,277	76,522	2,672,522	NA	3,294,040
2009	240,129	122,599	77,353	2,811,639	NA	3,251,720
2011	286,911	143,286	90,703	3,022,746	NA	3,543,646
2012	413,911	124,314	140,435	2,010,585	NA	2,689,245
2013	363,742	187,682	148,319	3,169,355	NA	3,869,098
TOTALS	13,833,167	2,419,107	2,672,484	66,686,384	NA	93,993,312[5]

[1] All 1979 and 1980 landings, and Texas landings from 1979 through 1997, rounded off to nearest 1,000; from 2000 Red Drum Stock Assessment, NMFS. Landings from 1981 through 2013, for Louisiana, Mississippi, Alabama and the west coast of Florida from National Marine Fisheries Service landings data.
[2] West coast only
[3] Excludes Texas landings from 1998 onward
[4] Not Available from NMFS
[5] Excludes Texas landings from 1998 onward

TABLE 5

NUMBER OF **RECREATIONAL** FISHING TRIPS
IN GULF OF MEXICO
FOR ALL SPECIES 1981-2013[1]

YEAR	Florida[2]	Alabama	Mississippi	Louisiana	Texas	TOTAL[3]
1981	9,520,481	523,318	662,224	1,358,307	NA	12,064,330
1982	8,807,769	1,362,914	717,317	2,535,776	NA	13,423,776
1983	14,521,381	1,739,128	1,037,527	2,685,030	NA	19,983,066
1984	16,521,499	612,828	797,341	1,711,053	NA	19,642,721
1985	11,583,246	710,846	572,086	2,553,624	NA	15,419,802
1986	14,367,176	866,722	776,626	3,029,420	NA	19,039,944
1987	12,321,111	622,080	775,582	2,370,674	NA	16,089,447
1988	14,730,478	1,182,515	907,695	2,922,611	NA	19,743,299
1989	12,031,576	622,719	704,496	2,263,719	NA	15,622,510
1990	9,922,602	722,805	686,439	1,978,380	NA	13,310,226
1991	14,261,115	648,774	843,905	2,419,805	NA	18,173,599
1992	13,763,989	763,018	1,001,436	2,550,806	NA	18,079,249
1993	12,928,092	933,061	866,103	2,703,754	NA	17,431,010
1994	13,166,982	886,949	964,498	2,485,308	NA	17,503,737
1995	12,396,870	998,539	1,053,434	2,941,473	NA	17,390,316
1996	12,331,873	931,884	945,154	2,823,868	NA	17,032,779
1997	13,384,436	1,024,177	999,093	3,185,378	NA	18,593,084
1998	12,234,580	968,485	827,536	2,672,764	NA	16,703,365
1999	11,296,851	1,169,914	805,518	2,621,446	NA	15,893,729
2000	15,086,213	1,086,818	1,093,144	3,751,609	NA	21,017,784
2001	16,388,611	1,635,798	1,250,045	3,615,244	NA	22,889,698
2002	14,418,275	1,190,004	1,038,353	3,018,946	NA	19,665,578
2003	16,008,974	1,499,989	1,176,788	4,270,921	NA	22,956,672
2004	17,795,711	2,250,691	1,179,292	5,203,514	NA	26,429,208
2005	16,694,805	1,604,207	925,717	4,065,078	NA	23,289,807

TABLE 5, CONTINUED

NUMBER OF **RECREATIONAL** FISHING TRIPS IN GULF OF MEXICO FOR ALL SPECIES 1981-2013[1]

YEAR	Florida[2]	Alabama	Mississippi	Louisiana	Texas	TOTAL[3]
2006	16,667,410	1,938,270	923,967	3,763,274	NA	23,292,921
2007	16,935,514	1,961,012	1,204,457	4,188,282	NA	24,289,265
2008	17,497,165	1,703,946	968,686	4,620,056	NA	24,789,853
2009	15,677,320	1,712,587	1,079,328	4,128,014	NA	22,597,249
2010	14,266,196	1,686,157	1,232,593	3,862,487	NA	21,047,433
2011	13,900,677	2,483,465	1,615,390	4,576,247	NA	22,575,779
2012	14,780,184	2,305,286	1,950,449	4,136,564	NA	23,172,483
2013	15,949,030	2,862,430	1,760,758	4,661,154	NA	25,233,372
TOTALS	462,158,192	43,211,336	33,342,977	105,674,586	NA	644,387,091

[1]Includes trips in both inshore and offshore waters for all species, not just redfish. Data from the NMFS Marine Recreational Fishing Survey
[2]West coast only
[3]For four Gulf Coast states, excluding Texas

BIBLIOGRAPHY

Since I was a net fisherman at the time, I personally witnessed many of the occurrences recounted in this book. Additionally, the thorough reporting by both the New Orleans *Times-Picayune* and Baton Rouge's *The Advocate* provided a historical record of that period that was more than helpful in organizing the chronology of events. Pertinent articles in both newspapers are too numerous to cite.

Unless otherwise noted, all sport and commercial fishery landings data cited in the text and tables are from the National Marine Fisheries Service, Fisheries Statistics Division. The two website addresses where such data can be obtained are listed below under that agency's name.

Other references include:

Adkins, B. Gerald. Shrimp with a Chinese flavor. *Louisiana Conservationist*, Vol. 25, No. 7-8. 1973.

_____; Tarver, Johnnie; Bowman, Philip; and Savoie, Brandt. A study of the Commercial Finfish in Coastal Louisiana. Louisiana Department of Wildlife and Fisheries, Seafood Division. Technical Bulletin No. 29. 1979.

_____ and M. Bourgeois. An evaluation of gill nets of various mesh sizes. Louisiana Department of Wildlife and Fisheries, Technical Bulletin Number 30. 1982.

_____, et. al. A creel survey of Louisiana recreational saltwater anglers. Louisiana Department of Wildlife and Fisheries, Technical Bulletin 41. 1990.

_____, et al. A Biological and Fisheries Profile of Southern Flounder, *Paralichthys lethostigma*, in Louisiana. Fisheries

Management Series No. 6, Part 1. 1996.

Anonymous. "The Original Picayune Creole Cook Book." Times-Picayune Publishing Corporation, New Orleans, Louisiana. 1901, 1906, 1916, 1922, 1928, 1936, 1938, 1942, 1945, 1947, 1954, 1966, 1971.

Barras, John A. Land area change in coastal Louisiana after the 2005 hurricanes—a series of three maps; U.S. Geological Survey Open-File Report 06-1274. 2006.

_____. Land Area Change and Overview of Hurricane Impacts in Coastal Louisiana, 2004-08: U.S. Geological Survey Scientific Investigations Map 3080, Scale 1:250,000, 6 p. Pamphlet, http://pubs.usgs.gov/sim/3080/. 2009.

_____, J.C. Bernier, and R.A. Morton. Land Area Change in Coastal Louisiana—A Multidecadal Perspective (from 1956 to 2006): U.S. Geological Survey Scientific Investigations Map 3019, scale 1:250,000, 14p. Pamphlet, http://pubs.usgs.gov/sim/3019/. 2008.

Berry, F.H. and Eversen, E.S. Pompano: biology, fisheries, and farming potential. Proc. Gulf Caribb. Fish. Inst. 19:116-128. 1967.

Bertrand, A.L. Marine recreational finfishermen in Louisiana: Socioeconomic study of licensed recreational finfishermen fishing in Coastal Study Area IV. Louisiana State University, Coastal Ecology and Fisheries Institute, Technical Series 3. 1984.

Bharadwaj, Latika, et al. 2012. Louisiana Commercial Saltwater Finfish Dealers: Analyzing Point of First Sales Data for the Louisiana Commercial Finfish Sector: 2000-2009 with an Examination of Changes in First Receivers' Activities after Hurricanes Katrina and Rita in 2005 and Gustav & Ike in 2008.

BIBLIOGRAPHY

Louisiana Department of Wildlife and Fisheries, Office of Fisheries, Socioeconomic Research and Development Section, Baton Rouge, LA. August, 2012.

Blankenship, Karl. Atlantic sturgeon back in Bay, or did they ever leave? *Chesapeake Bay Journal*, Vol. 24, No. 9, pp. 1, 18-19, December, 2014.

Boesch, Donald, et al. Scientific Assessment of Coastal Land Loss, Restoration and Management in Louisiana. Journal of Coastal Research, Special Issue No. 20. 1994.

Bourgeois, Martin J. et al. A Fisheries Management Plan for Louisiana Spotted Seatrout, *Cynoscion nebulosus*. Fisheries Management Plan Series No. 3, Part 1. 1995.

Bowman, Philip; Adkins, Gerald; and Tarver, Johnnie. A Profile of the Commercial Finfishermen in Coastal Louisiana. Louisiana Department of Wildlife and Fisheries, Seafood Division. Technical Bulletin No. 25. 1977.

Bridges, Tyler. "Bad Bet on the Bayou: The Rise of Gambling in Louisiana and the Fall of Governor Edwin Edwards." Farrar, Straus and Giroux, New York, NY. 2001.

Couvillion, B.R., J. A. Barras, G. D. Steyer, W. Sleavin William, M. Fisher, H. Beck, N. Trahan, B. Griffin, and D. Heckman. Land Area Change in Coastal Louisiana from 1932 to 2010: U.S. Geological Survey Scientific Investigations Map 3164, scale 1:265,000, 12 p. Pamphlet.
http://pubs.usgs.gov/sim/3164/downloads/SIM3164_Pamphlet.pdf. 2011.

Davis, D.W. Living on the edge: Louisiana's marsh, estuary and barrier island population. Transactions of the Gulf Coast

Association of Geological Societies, pp. 147-159. 1990.

Deegan, L.A., et al. Relationships among physical characteristics, vegetation distribution and fisheries yield in Gulf of Mexico estuaries. In: D. Wolfe (editor), "Estuarine Variability." Academic Press, New York, New York. p. 83-100. 1986.

Evans, Oliver. Melting pot in the bayous. American Heritage, Vol. 15, No. 1. 1963.

Fritchey, Robert. "Wetland Riders." New Moon Press, Golden Meadow, LA. 1993.

_____. Pompano Runnin' (And Jumpin' And Skippin'): For netters a small slice of Louisiana's coast is the whole pie. *National Fisherman*, February, 1999.

Gagliano, S.M. and van Beek, J.L. Geologic and geomorphic aspects of deltaic processes, Mississippi delta system. Hydrologic and geologic studies of coastal Louisiana, Report 1. Louisiana State University, Center for Wetland Resources, Baton Rouge, Louisiana. 1970.

Gary, D.L., and Davis, D.W. Recreational dwellings in the Louisiana coastal marsh. Sea Grant Publication LSU-T-79-002, Baton Rouge, Louisiana State University, Center for Wetland Resources. 1979.

Goodyear, C.P. Status of the red drum stocks of the Gulf of Mexico. Report for 1989. NMFS-SEFC, Miami Laboratory. CRD 88. 1989.

Gresham, Claude "Grits." "Fishes and Fishing in Louisiana." State of Louisiana, Department of Conservation Bulletin No. 23. Reprint by Claitor's Book Store, Baton Rouge, LA. 1965.

BIBLIOGRAPHY

Gulf South Research Institute. Louisiana Outdoor Recreation Survey. Louisiana Department of Culture, Recreation and Tourism, Baton Rouge. 1986.

Herke, W.H. Use of natural, and semi-impounded, Louisiana tidal marshes as nurseries for fishes and crustaceans. Louisiana State University, Baton Rouge, Louisiana. 1971.

_____ and B.D. Rogers. 1989. Threats to coastal fisheries. In: W.G. Duffy and D. Clark (editors), "Marsh Management in Coastal Louisiana: Effects and Issues—Proceedings of a Symposium." U.S. Fish and Wildlife Service Biological Report 89(22):196-212.

Hoese, Dickson H. & Moore, Richard H. "Fishes of the Gulf of Mexico, Texas, Louisiana and Adjacent Waters." Texas A&M Press. 1977.

Hoese, D. et al. A Biological and Fisheries Profile of Louisiana Red Drum, *Sciaenops ocellatus*. Louisiana Department of Wildlife and Fisheries, Fishery Management Plan Series, No. 4, Part 1. 1991.

Horst, Jerald. "Louisiana Seafood Products Handbook. A Reference Guide to Species, Availability, Product Forms, and Fisheries Facts." Louisiana Seafood Promotion & Marketing Board, 1998.

_____ & H. Holloway. Louisiana License Statistics and Trends, 1987-2000: Commercial Fishing, Recreational Gear, Commercial Wildlife, and Related Industries. Louisiana Sea Grant College Program and Louisiana State University. http://www.laseagrant.org. 2002.

Kearney/Centaur, Inc. Economic Impact of the Commercial Fishing Industry in the Gulf of Mexico and South Atlantic Regions. Gulf and South Atlantic Fisheries Development Foundation, Inc. Tampa, FL. 1984.

Kelso, W.E., et al. 1990 Survey of Louisiana Sport Fishermen. Louisiana State University Agricultural Center and Louisiana Department of Wildlife and Fisheries. Baton Rouge, Louisiana. 1991.

_____, et al. Survey of Louisiana Recreational Anglers 1991. Louisiana State University Agricultural Center and Louisiana Department of Wildlife and Fisheries. Baton Rouge, Louisiana. 1992.

_____, et al. 1993 Survey of Louisiana saltwater anglers. Louisiana Agricultural Experiment Station Mimeo Report No. 1994.

Leard, Richard, et al. The Striped Mullet Fishery of the Gulf of Mexico, United States: A Regional Management Plan. Gulf States Marine Fisheries Commission. Ocean Springs, MS. 1995.

_____, et al. The Menhaden Fishery of the Gulf of Mexico, United States: A Regional Management Plan. Gulf States Marine Fisheries Commission. Ocean Springs, MS. 1995.

Louisiana Department of Wildlife & Fisheries, Licensing data: http://www.wlf.louisiana.gov/licenses/statistics

Louisiana Department of Wildlife and Fisheries. Spotted Seatrout and Red Drum—An Overview. A Joint Fisheries/Seafood Division Task Force. January, 1983.

_____. A Stock Assessment for Louisiana Red Drum. Fisheries Management Plan Series, No. 4. 1991.

_____. A Biological and Fisheries Profile of Sheepshead, *Archosargus probatocephalus*, in Louisiana. Fisheries Management Plan Series No. 7, Part 1. 1996.

BIBLIOGRAPHY

Love, T.D. Survey of the sun-dried shrimp industry of the north central Gulf of Mexico. Commercial Fisheries Review, Vol. 29, No. 4. 1967.

Luquet, Clarence, Jr.; Roussel, John; Shepard, Joseph; and Blanchet, Harry. Black Drum Management Plan. Louisiana Department of Wildlife and Fisheries. May, 1990.

_____, et al. A Biological and Fisheries Profile for Black Drum, *Pogonias cromis,* in Louisiana. Louisiana Department of Wildlife and Fisheries. 1996.

Maddux, H.R., et al. Trends in finfish landings by sport-boat fishermen in Texas marine waters May 1974-May 1988. Texas Parks and Wildlife Department, Management Data Series Number 8. 1989.

Mapes, Karl A. et al. A Biological and Fisheries Profile of Striped Mullet, *Mugil cephalus,* in Louisiana. LDWF Fisheries Management Plan Series No. 5, Part 1. 1996.

Middleton, Leslie. Sturgeon study on tidal James offers evidence of fall spawning. *The Bay Journal,* October 29. 2013. http://www.bayjournal.com/article/sturgeon_study_on_tidal_james_offers_evidence_of_fall_spawn

National Marine Fisheries Service. Recreational Fishing Landings Data: http://www.st.nmfs.noaa.gov/recreational-fisheries/index

_____. Commercial Fishing Landings Data: http://www.st.nmfs.noaa.gov/commercial-fisheries/index

Nixon, S. Between coastal marshes and coastal waters—a review of twenty years of speculation and research on the role of salt marshes in estuarine productivity and water chemistry. pp. 437-525. In: P.

Hamilton and K. MacDonald (editors), "Estuarine and Wetland Processes." Plenum Publishing Corp., New York, New York. 1980.

Penland, S., et al. Coastal Land Loss in Louisiana. Gulf Coast Association of Geological Societies Transactions, Vol 40. 1990.

Ridgeway, James. Environmental Espionage: Inside a Chemical Company's Louisiana Spy Op. www.motherjones.com, May 20, 2008.

Roberts, Kenneth J. A synopsis of economic impacts associated with recreational and commercial use of Louisiana seafoods (Splitting Atoms is Easier Than Dividing Fish). Louisiana Cooperative Extension Service, Sea Grant College Program, Louisiana State University, Baton Rouge, LA. 1986.

_____. Allocating Spotted Seatrout. Louisiana Sea Grant Program, Louisiana State University, Baton Rouge, LA. 1991.

_____, Horst, Jerald W.; Roussel, John E.; and Shephard, Joseph A. "Defining Fisheries." Louisiana Sea Grant Program, Louisiana State University, Baton Rouge, LA. 1991.

Russell, S.J. et al. State/Federal Cooperative Fishery Statistical Program in Louisiana. Annual Report 1985-86. Louisiana State University, Coastal Fisheries Institute, Publ. No. LSU-CFI-86-11. 1986.

_____. State/Federal Cooperative Fishery Statistical Program in Louisiana. Annual Report 1986-87. Louisiana State University, Coastal Fisheries Institute, Publ. No. LSU-CFI-87-12. 1987.

Southwick Associates, Inc. The Economic Benefits of Fisheries, Wildlife and Boating Resources in the State of Louisiana. Prepared for the Louisiana Department of Wildlife and Fisheries. 1997.

BIBLIOGRAPHY

Sterns, Silas. Fisheries of the Gulf of Mexico: The fishery interests of Louisiana, in Fisheries and Fishery Industries of the United States. Washington, D.C., U.S. Government Printing Office. 1887.

Thompson, Bruce. Life history and population dynamics of commercially harvested striped mullet *Mugil cephalus* in coastal Louisiana. Final Report Louisiana Bd. Regents. Coastal Fisheries Institute. LSU-CFI-90-01. 1989.

Titre, J.P., et al. Valuing wetland recreational activities on the Louisiana coast: Final report to U.S. Army Corps of Engineers, New Orleans District. 1988.

Turner, R.E. Louisiana's coastal fisheries and changing environmental conditions. pp. 363-367. In: J.W. Day, D.D. Culley, Jr., R.E. Turner, and A.J. Mumphrey, Jr. (editors), Proceedings: Third Coastal Marsh and Estuary Management Symposium. Louisiana State University, Division of Continuing Education, Baton Rouge. 1979.

_____ & Boesch, Donald F. Aquatic Animal Production and Wetland Relationships: Insights Gleaned Following Wetland Loss or Gain, in "The Ecology and Management of Wetlands, Volume 1: Ecology of Wetlands." Timber Press, Portland, OR. 1988.

U.S. Department of Commerce. 1975 National survey of hunting, fishing, and wildlife associated recreation. U.S. Fish and Wildlife Service, Washington, D.C. 1977.

U.S. Department of the Interior, Fish and Wildlife Service. 1985 National survey of fishing, hunting, and wildlife associated recreation. Louisiana. U.S. Government Printing Office, Washington, D.C. 1988.

_____ and U.S. Department of Commerce Bureau of the Census. 1991 National survey of fishing, hunting, and wildlife-associated recreation. U.S. Government Printing Office, Washington, D.C. 1993.

Walker, Michael, Editor. "Sport Fishing USA." The U.S. Department of the Interior, Bureau of Sport Fisheries and Wildlife, Fish and Wildlife Service. U.S. Government Printing Office, 0-441-620, 1971.

Wallace, Richard K.; Hosking, William; and Szedlmayer, Stephen T. Fisheries Management for Fishermen: A manual for helping fishermen understand the federal management process. Auburn Univeristy Marine Extension & Research Center, Sea Grant Extension, Mobile, AL. 1994.

Williams, S. Jeffress, Penland, Shea, and Sallenger, Jr., Asbury H. Atlas of Shoreline Changes in Louisiana from 1853 to 1989. U.S. Geological Survey, Miscellaneous Investigations Series I-2150-A. 1992.

Wilson, C.A. et al. The age structure and reproductive biology of sheepshead, *Archosargus probatocephalus*, landed in Louisiana. Final report to the Board of Regents. Coastal Fisheries Institute, Louisiana State University, Baton Rouge, LA. 1989.

•••••••••
CLIFF GLOCKNER LACOMBE

LSU Agricultural Center. Louisiana Cooperative Extension Service. Louisiana Summary: Agriculture and Natural Resources. 1994.

_____. Louisiana Cooperative Extension Service. Louisiana Summary: Agriculture and Natural Resources. 1997.

BIBLIOGRAPHY

KERRY LEBAUVE COCODRIE

Cobb, Thomas Blum, and Currie, Mara. "Images of America: Houma." Arcadia Publishing, Charleston SC; Chicago IL; Portsmouth, NH; San Francisco, CA. 2004.

ROBERT FRITCHEY GOLDEN MEADOW & LEEVILLE

Fritchey, Robert. Taken By Storm. *National Fisherman*, January, 2000.

An Economic Survey of Lafourche Parish. Division of Research, College of Commerce, LSU, May 1949.

GENERAL

Alvarez, Leilani. Conflicts in Fisheries Management: Commercial Fishers and Managers Speak Out in Tampa Bay, Florida, USA and Moreton Bay, Queensland, AUS. B.S. Thesis, Department of Biology, Emory University. 1996.

Anderson, L.G. Necessary components of economic surplus in fisheries economics. Can. J. Fish. Aq. Sci. Vol. 37, No. 5, pp. 858-870. 1980.

Brandt, Andres von. "Fish Catching Methods of the World." Fishing News (Books) Ltd., 110 Fleet St., London EC4. 1964.

Carey, Richard Adams. "Against the Tide: The Fate of the New England Fisherman." Houghton Mifflin Co., New York, NY. 2000.

Culliton, T.J., Warren, M.A., et al. Fifty Years of Population Change

Along the Nation's Coasts, 1960-2010. National Oceanic and Atmospheric Administration Coastal Trends Series, Second report. Rockville, MD. 1990.

DesJardins, Marc. El Niño is Back: Pacific water temperatures are on the rise. *National Fisherman,* October 1997.

Easley, J.E., V.K. Smith, M.K. Wohlgenant and W.N. Thurman. Allocating Recreational-Commercial Fishery Harvests: Literature Reviews and Preliminary Work Toward Modeling the Issue. Gulf and South Atlantic Fisheries Development Foundation, Inc., Tampa, FL. March 1989.

Edwards, Steven. An Economics Guide to Allocation of Fish Stocks between Commercial and Recreational Fisheries. NOAA Technical Report NMFS 94, 1990.

Freese, Curtis H., Editor. "Harvesting Wild Species: Implications for Biodiversity Conservation." The Johns Hopkins University Press, Baltimore, MD. 1997

Fritchey, Robert. Anatomy of a well-oiled campaign to Ban the Nets! *National Fisherman,* December, 1994.

_____. Southern Netters Fight Campaigns to Ban Their Gear. *National Fisherman,* July 1995.

Galtsoff, Paul S., editor. "Gulf of Mexico: Its Origin, Waters, and Marine Life." Fishery Bulletin of the Fish and Wildlife Service, Vol. 55, U.S. Government Printing Office, Washington, DC. 1954.

Gay, Joel. Crude Behavior: With new evidence it says points to a mistrial and an appeal coming up this summer, Exxon is set to add insult to Alaska fishermen's injury. *National Fisherman,* July 1998.

BIBLIOGRAPHY

Grover, J.H. (ed.). Allocation of Fishery Resources. UN Food and Agricultural Organization. Auburn University, Alabama. April, 1980.

Hoffer, Eric. "The True Believer: Thoughts on the Nature of Mass Movements." HarperCollins Publishers, Inc., New York, NY. 1951.

Hopkins, S.H. and Petrocelli, S.R. Limiting factors affecting the commercial fisheries in the Gulf of Mexico. Estuarine Pollution Control & Assessment. Proceedings of a Conference, Vol. 1, US EPA, Office of Water Planning & Standards, Washington, D.C. 1975.

Johns, M.A. Trends in Texas commercial fishery landings, 1972-1989. Texas Parks and Wildlife Department, Management Data Series Number 37. 1990.

Jordan, D.S. and B.W. Evermann. Fishes of North and Middle America. Smithsonian Institution. Bulletin No. 47, Part II. 1896. (Reprinted by T.F.H. Publications, 1963).

Kronman, Mick. They're Baaack!: Sardines, the fish that built Cannery Row, may be swarming the coastline, but the F/V Aliotti Bros. still has to find them. *National Fisherman,* March 1999.

McClane, A.J., editor. "McClane's New Standard Fishing Encyclopedia and International Angling Guide." Holt, Rinehart and Winston, New York, NY. 1974.

National Marine Fisheries Service. The 1976 Magnuson Fishery Conservation and Management Act, as amended through 1996. U.S. Department of Commerce, National Oceanic and Atmospheric Administration, 1996.

Ruello, N.V. and Henry, G.W. Conflict Between Commercial and Amateur Fishermen." *Australian Fisheries.* March, 1977.

Smith, S. and Jepson, M. Big fish, little fish: politics and power in the regulation of Florida's marine resources. *Social Problems* 40(3): 301-312. 1993.

_____. Social implications of changes in fisheries regulations for commercial fishing families. *Fisheries* 20(7):24-26. 1995.

Thunberg, E.M. et al. Social and economic issues in marine fisheries allocations: a Florida perspective. *Trends* 31(1):31-36. 1994.

Zile, Dexter Van. Hot Air Over Water: Climate change and its causes are arguable, but there's no escaping the connection between ocean conditions and fish. *National Fisherman*, August, 2001.